TROUBLESHOOTING ELECTRONIC EQUIPMENT WITHOUT SERVICE DATA

TROUBLE-SHOOTING ELECTRONIC EQUIPMENT WITHOUT SERVICE DATA

Second Edition

Robert G. Middleton

Prentice Hall
Englewood Cliffs, New Jersey, 07632

Library of Congress Cataloging-in-Publication Data

Middleton, Robert Gordon
   Troubleshooting electronic equipment without service data / Robert
G. Middleton.—2nd ed.
    p. cm.
   Includes index.
   ISBN 0-13-931164-5
   1. Electronic apparatus and appliances—Maintenance and repair.
I. Title.
   TK7870.2.M535   1989
   621.381—dc19
                                         88-32132
                                            CIP

Editorial/production supervision and interior design: Eve Mossman
   & Jacqueline A. Jeglinski
Cover design: George Cornell
Manufacturing buyer: Mary Ann Gloriande

 © 1989 by Prentice-Hall, Inc.
A division of Simon & Schuster
Englewood Cliffs, New Jersey 07632

The publisher offers discounts on this book when ordered
in bulk quantities. For more information, write:
       Special Sales/College Marketing
       Prentice-Hall, Inc.
       College Technical and Reference Division
       Englewood Cliffs, NJ 07632

Printed in the United States of America
10 9 8 7 6 5 4 3 2

ISBN 0-13-931164-5

Prentice-Hall International (UK) Limited, *London*
Prentice-Hall of Australia Pty. Limited, *Sydney*
Prentice-Hall Canada Inc., *Toronto*
Prentice-Hall Hispanoamericana, S.A., *Mexico*
Prentice-Hall of India Private Limited, *New Delhi*
Prentice-Hall of Japan, Inc., *Tokyo*
Simon & Schuster Asia Pte. Ltd., *Singapore*
Editora Prentice-Hall do Brasil, Ltda., *Rio de Janeiro*

CONTENTS

3 PROGRESSIVE AUDIO TROUBLESHOOTING TECHNIQUES   59

4 TROUBLESHOOTING RADIO RECEIVERS WITHOUT
SERVICE DATA   85

5 ADDITIONAL RADIO TROUBLESHOOTING TECHNIQUES   113

6 PROGRESSIVE RADIO TROUBLESHOOTING TECHNIQUES 143

7 TELEVISION TROUBLESHOOTING WITHOUT SERVICE DATA 163

8 ADDITIONAL TELEVISION TROUBLESHOOTING METHODS 180

9 PROGRESSIVE TELEVISION TROUBLESHOOTING METHODS   194

10 FOLLOW-UP TELEVISION TROUBLESHOOTING METHODS   211

11 COLOR CIRCUIT TROUBLESHOOTING   228

12 TAPE RECORDER TROUBLESHOOTING   254

13 CLOSED CIRCUIT TELEVISION CAMERA TROUBLESHOOTING   273

A INTERNATIONAL FREQUENCY ALLOCATIONS   299

INDEX   307

A WORD
FROM THE AUTHOR

Today, as in the past, meeting the challenge of troubleshooting audio, radio, television, and related electronic equipment without service data confronts us. This book will help you meet that challenge.

Experienced troubleshooters know that service data may not be available for various reasons:

1. The manufacturer sometimes releases a new product with the intention of following up with service data when the factory is less rushed.
2. It may be difficult to run down the address of a foreign consumer electronics product manufacturer.
3. There is always some time lag involved in the availability of service data from professional publishers.
4. Service data may never be issued for various proprietary brands of electronic equipment.
5. Numerous design changes are sometimes made for a particular production run, which make the available service data misleading.
6. Although service data may be available, the troubleshooter may decide that it doesn't pay to invest ten dollars for a small-ticket job.
7. Sometimes the available service data is so sketchy that it is of no real assistance in the troubleshooting procedure.
8. After a unit of electronic equipment has been serviced numerous times in various shops, troubleshooters may be confronted with one or more

nonstandard repairs with which they must contend, or they must return the unit to their customers on a concession loss basis.

9. Manufacturers "goof" once in a while and issue service data with technical errors, drafting errors, or incorrect entries. The professional troubleshooter needs to have the ability to spot such errors.

10. Service-data writers sometimes assume troubleshooters have sophisticated (and expensive) test equipment on hand, which, in fact, they do not. In turn, troubleshooters are thrown upon their own resources, although service data are "available."

As in the previous edition, many never-before-published troubleshooting techniques are explained in this book, with case histories to show how the new techniques are applied and evaluated in actual practice, step by step. Included among these unique troubleshooting techniques are:

Voltage-forcing in-circuit tests

The sound of troubles

Quick-checking FM converter operation with a radio receiver

Signal-killer quick checker

Neon-bulb oscillating field-intensity checker

AC/DC current quick checker

Audio troubleshooting phase tests

The sound of phase shift

AM detector "spotter"

FM detector "spotter"

Incidental amplitude modulation tests

Recording DC voltage monitor

The sound of full rectification

The sound of overmodulation

Identification of 60-Hz buzz

Thread test for high-voltage DC

Chroma signal checking with radio receiver

Quick check for localization of intercarrier IF section

An important feature of this new edition is its coverage of integrated circuits from the troubleshooting viewpoint. That is, a transistor is a single-function device, whereas an integrated circuit is a multifunction device. In turn, the troubleshooter must now make multifunctional analyses of trouble symptoms when integrated circuitry is of concern.

Phase checks are among the innovative audio troubleshooting procedures described in this new edition. Troubleshooters ordinarily disregard phase relations because the ear is virtually unresponsive to phase shifts in an audio signal. However, phase quick-checks can be quite informative in preliminary troubleshooting procedures and merit consideration. Phase distortion is the result of unequal reproduction of the various phase relationships in a signal (particularly unequal reproduction of voltage-current phases).

Ideally, practically all signal voltages in an audio system are either precisely in-phase or precisely 180 degrees out of phase with one another. In the event of malfunction, ideal relations "go by the board." For example, if a coupling capacitor or a bypass capacitor deteriorates, it generally introduces phase shift into the signal, even at mid-band frequencies. In turn, a simple phase quick-checker will localize the trouble area without any analysis of circuit details or circuit action.

Of course, "the exception proves the rule," and there is an exception to the rule above. In the case of tone-control circuits, phase shift is inherent in the circuit action, and this circumstance must be kept in mind. Fortunately, tone-control circuitry is plainly evident in audio systems, and is easily recognized when troubleshooting without service data.

Five chapters deal with TV troubleshooting without service data. Preliminary approaches are explained, with notes on quick checks such as temperature tests that require no knowledge of the circuit under test, nor its function. PC board markings are discussed along with general "landmarks." Typical causes of common trouble symptoms are noted, along with preliminary test procedures. Sneaky malfunctions caused by "garbage" in the V_{cc} line are described. It is a maxim in troubleshooting without service data that "you can't know too much about circuit action." A new section in the text gives a quick run-down on the circuit actions of major importance in TV receivers.

Although the oscilloscope is discussed fully, other instruments such as the DVM are stressed in test procedures. Considerable attention is also given to shop-constructed testers and their application because many troubleshooters do not have a scope available, and some scope owners do not feel "comfortable" with this comparatively complex instrument. The text explains how to make the most out of simple and inexpensive test equipment, along with some unexpected tricks of the trade.

Tape recorders are in wide use, and a chapter is included on basic mechanical difficulties, quick checks, variational checking, and how to cope with audio system interference. An easy and informative recycling quick check is explained, with evaluation of test results.

Troubleshooting without service data is both an art and a science. The troubleshooter's mental attitude is as important as his or her technical know-how. Technicians who are comparatively experienced in conventional troubleshooting procedures sometimes have a mental block when they are confronted by a strange unit of electronic equipment—without any service data.

This "mental block" occurs because the *fundamental approach* to trouble symptom analysis and testing procedure is different when no service data are at hand.

There are various "degrees" of difficulty that are encountered at the outset when a strange unit of electronic equipment is confronted. For example, many consumer products have a small circuit diagram inside the cabinet or case. (The operator's manual may also include a circuit diagram.) Sometimes the various functional sections are identified on the component side of the circuit board. Progressive parts numbers may be provided although functional sections are not identified. In any situation, resistors are generally color-coded, and most of the capacitors will have their values marked.

Most service shops keep a library of service data, and the general rule for technicians is to read the manual. However, many old-timers seldom do this. Over the years, they have come to know that all electronic equipment of a certain type has to work in a certain way, and that there are specific trouble symptoms for each section of the circuitry.

Furthermore, each manufacturer has his or her own particular style of products so that there is a resemblance among them. It is therefore quite feasible for the experienced troubleshooter to operate without service data, using methods similar to those described here. This does not imply that service data should be ignored when available, but the experienced technician can generally manage without it.

It is apparent that the person who is in the ideal position to troubleshoot a unit of electronic equipment without service data is the engineer who designed the unit, and who perhaps also wrote the service manual for the unit. In other words, the engineer is completely familiar with the circuitry and with the circuit actions, plus the production test procedures that are used to determine whether the unit is operating within acceptable limits. That is, "you can't know too much about circuit action when troubleshooting without service data."

In this book, various tricks of the trade are described. Many of these "tricks" require both a thorough knowledge of the way circuits operate and sufficient experience to recognize when test-equipment readings are "in the ballpark." It is in the service shop that similar equipment in working order will most likely be available for comparison tests. In the home, comparison tests will generally be confined to stereo systems, various intercom arrangements, and walkie-talkies. In turn, more generalized approaches that do not employ comparison tests must often be followed.

Just as comprehensive knowledge of circuit action is very desirable when troubleshooting without service data, so is extensive experience in circuit testing. Here, the beginner is at a disadvantage, inasmuch as he or she must "start from scratch." It may be observed that more rapid and easier progress can be made in acquiring experience in circuit testing if the beginner takes time to construct and check out various basic kit equipment, such as an AM radio kit. (There is no substitute for "hands-on" experience.)

The same observation applies to some test equipment, and this book may have somewhat more extensive application for service technicians than for hobbyists and experimenters. However, this book is for the hobbyist or experimenter who is concerned with electronic circuitry and interested in many of the test procedures and home-made test equipment units featured in the text. Again the beginner is well advised to take time to construct and check out various basic kit test equipment, such as a 1000 ohms-per-volt multimeter.

Various facts that are taken for granted in conventional troubleshooting procedures may not be immediately apparent when no service data are at hand. For example, in the absence of service data, the technician cannot refer to a layout diagram and immediately point to an RF section, an IF section, a video section, a sync section, an audio section, and so on. He or she cannot point immediately to an intercarrier-sound transistor, an AGC transistor, or even to the emitter, base, and collector terminals of various transistors.

Instead, the troubleshooter must make a visual inspection of the circuit board and follow some preliminary map-out procedures before he or she can say what a particular transistor, diode, or integrated circuit is supposed to do. This is an approach that is quite different from looking at a service manual in which every component, device, functional section, and interconnection detail is set forth, with operating voltages, circuit resistances, and scope waveforms.

To get the feel of troubleshooting without service data, the technician must depart from his or her established perspective, and start with a "black-box" viewpoint—that is, elementary questions, that ordinarily have obvious answers, must now be checked out by specific tests and measurements, some of which are new and unexpected. There are explanations throughout on how you can build your own specialized testers for troubleshooting without service data.

Virtually all electronic troubleshooters agree on the basic principles that time is money and that knowledge is power. Your success in the profession of electronic troubleshooting is limited only by the horizons of your technical know-how. The novel tricks of the trade, new techniques, and troubleshooting approaches described and illustrated in this unusual book provide key stepping stones to take you from your present position to your goal.

Robert G. Middleton

1

AUDIO TROUBLE-SHOOTING TECHNIQUES

Overview • Preliminary Trouble Analysis • Second Step • Volt-Amp-Wattmeter • Audio Current "Sniffer" • "Ballpark" Temperatures • Integrated Circuits • New Test Equipment for Troubleshooting without Service Data • Resistor, Capacitor, Diode, and Transformer Temperatures • High-Speed Differential Temperature Checker • Capacitor Identification • Working with Capacitors • Follow-up DC Voltage Measurements • Finger Test • Diode Identification; Quick Checks • Field-Effect Transistors • Parts Numbering Sequence • Measuring Inductance with a DVM • In-Circuit Transistor Testing • Functional Overview of Consumer-Electronic ICs • Oscilloscope Capabilities

OVERVIEW

Troubleshooting electronic equipment without service data requires one of two basic approaches, depending on whether the equipment is workable (although malfunctioning), or whether it is unworkable (dead). When a unit is workable, although malfunctioning, the preliminary analysis consists in "sizing up" the trouble symptoms for clues concerning the nature of the fault. In the case of an audio preamplifier unit, for example, the sound output should be analyzed with respect to each input port, such as tape, phono, tuner, and mic inputs. (See Table 1-1).

On the other hand, when a unit is unworkable (dead), the volume control of the preamp should be turned up to maximum, and the noise output, if any, analyzed for comparative level, characteristics, and response to control variation. For example, the noise level may be found to be excessively high, or almost inaudible. Its characteristics may consist of a steady hiss, or a hiss interrupted by crackling noises. Sometimes the noise output will consist primarily of a loud hum. In any case, these factors all provide clues concerning the nature of the fault.

SECOND STEP

In most cases, the second step consists of chosen measurements and quick checks in an attempt to narrow down the suspects. Transistors and integrated circuits provide helpful "landmarks" in preliminary troubleshooting procedures. As explained in greater detail on the following pages, DC voltage measurements at transistor or IC terminals can often provide helpful preliminary data in case the unit is functioning.

However, if the unit is dead with audible noise output, it is often expedient to proceed from the output port with a signal injector back toward the input port to narrow down the trouble area. This procedure requires determination of the stage sequence, as detailed on the following pages.

VOLT-AMP-WATTMETER

You may find the normal rated power consumption for the unit printed on the rear of the case. If so, a useful quick check can be made with a volt-amp-wattmeter. This is a tester that is plugged in series with the line cord. As an

---

Table 1-1

Preliminary Trouble Analysis of Faults in Audio Preamps

Unit Workable, with Malfunction	*Unit "Dead"*
* Is operation normal on one or more input ports (tape, phono, tuner, microphone)?	* Is unit "dead" only with respect to audio output, or is it completely dead (lack of any hiss or hum when volume control is advanced to maximum)?
* Stereo unit: Is operation normal on one channel (L or R)?	
* Nature of malfunction: Single trouble symptom, such as weak output, distorted output, hum, noise, howl? Combination trouble symptom, such as weak and noisy output, or distortion[a] with high hum level?	* Is there any "click" output when the power switch is turned on or off?
	* After complete warm-up, is the unit intermittent (are there momentary bursts of sound output, for example)?
* Is the trouble symptom steady or intermittent?	* If there is hiss, hum, or crackling noises, do the noise characteristics change during warm-up, or as the unit "cooks" for several minutes?
* Are there obvious clues, such as overheated components or devices, burnt odors, melted wax, audible mechanical hum, erratically operating controls?	
	* Is the unit workable for a few seconds during warm-up, and then suddenly goes "dead"?
* Does the circuit board show evidence of tampering?	* Does other equipment, such as an all-wave radio receiver in the vicinity of the unit develop interference symptoms (interference radiated by the "dead" preamp)?
* Is there evidence of mechanical violence, such as cracked circuit boards?	
* Does the circuit board have an appearance of operation in a damp location?	* Does the unit suddenly start to work when the power-supply voltage is slightly reduced?
* Have obviously nonstandard repairs been made on the unit?	* Is the unit susceptible to stray fields (for example, does it devel-

(continued)

Table 1-1 (*Continued*)

Unit Workable, with Malfunction	Unit "Dead"
* How does unit respond to tapping of individual components and devices?	op any sound output when placed near a CB transmitter, or a TV receiver)?
* How does the unit respond to moderate flexing of circuit board?	* Does an audio signal tracer show that any portion of the amplifier is workable?
* Does the trouble symptom change after the unit has warmed up completely?	* Is the power consumption normal?

ªSee Chapter 4 for discussion of "The Sound of Troubles."

illustration, the Realistic (Radio Shack) SA-102 stereo amplifier is rated for a power consumption of 25 watts. If, in turn, it is found that the unit is drawing 15 or 35 watts, the advisable procedure is to make quick checks with a temperature probe and DVM to determine where overheating may be occurring, or where subnormal operating temperatures may be occurring.

Preliminary trouble analysis is directed toward narrowing down the trouble area and pinpointing the fault to avoid a time-wasting "shotgun approach."

AUDIO CURRENT "SNIFFER"

One of the most useful time-savers that can be kept handy on the audio bench is the audio current "sniffer" shown in Figure 1-1(a–d). It is the simplest possible arrangement that can be used to detect the presence of audio current, and to trace audio current through a wiring system. The "sniffer" is surprisingly sensitive, and will clearly indicate the presence of comparatively small AC currents. Its maximum effectiveness occurs at about 500 Hz, although it reproduces higher and lower audio frequencies also. This one is hard to beat.

Figure 1-1

TRICK OF THE TRADE: In a preliminary troubleshooting checkout of an audio system, we often wish to quickly determine whether there is DC current flowing in a wire, and whether it may be accompanied by an AC audio current. It is easy to do this if we make a quick

Figure 1-1 (continued)

Printed-circuit (or other) conductor

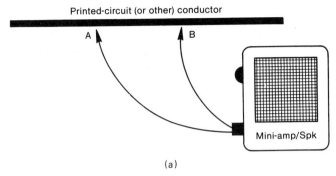

(a)

check with a mini-amp/speaker as shown above. The input test leads are touched at arbitrary points A and B along the conductor. *If there is a click from the speaker, there is DC current flow in the conductor—but if there is no click there is very little or no DC current flowing in the conductor. Next, if there is a tone from the speaker, there is AC audio current flow in the conductor—but if there is no tone there is very little or no AC audio current flowing in the conductor.* A Radio Shack 277-1008B mini-amplifier/speaker is very useful for this quick check. It provides a gain of about 1700 times, and develops maximum rated output with a l-mV input voltage change. When the volume control is advanced to maximum, even a comparatively small AC current flow in the conductor will produce sufficient IR drop along the conductor to provide a clearly audible tone from the speaker.

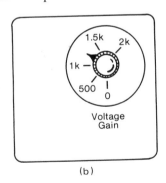

(b)

Note: This type of voltage-gain calibration scale can be provided easily on the back of the mini-amp/spk and considerably increases the usefulness of the tester. It shows the approximate gain of the tester at any setting of the volume control. (The original volume control is removed from the edge of the case, and is relocated on the back of the case and provided with a hand-calibrated scale. The scale is marked off with the aid of an audio oscillator and AC DVM.)

Figure 1-1 (continued)

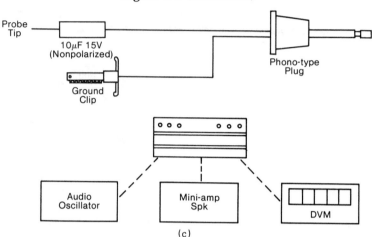

(c)

Note: This is an essential test-lead arrangement for the mini-amp/spk when signal tracing in active circuitry. It is an isolating probe that passes an AC signal and blocks drain-off of DC voltage from the circuit under test. Observe that a nonpolarized type of electrolytic capacitor must be used, since the polarity of the DC voltage that is present is usually unknown. If you do not have a 10 μF nonpolarized capacitor available, you can use two 20 μF polarized capacitors connected in series, back to back. Capacitors should be rated for at least 15 V.

Technical Note: As indicated in the diagram, the troubleshooter must often take a "black-box" approach when service data is not available. It is very helpful to have a comparison "black box" at hand for evaluation of DC voltage and resistance values. In any case, the mini-amp/spk serves to localize the trouble area, and the defective device or component can usually be pinpointed by follow-up voltage and resistance measurements.

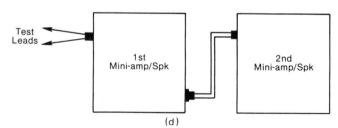

(d)

Note: This is an extra-high-gain audio signal-tracing arrangement that is very useful in some situations. For example, when tracing a sub-normally weak or very low-level signal, a single mini-amp/spk may

Figure 1-1 (continued)

reproduce a barely audible output. On the other hand, when two mini-amp/spk units are operated in series, the barely audible output will become plainly audible. Observe that the input jack of the second mini-amp/spk is connected to the earphone jack of the first mini-amp/spk. In turn, there is sound output only from the second mini-amp/spk.

The inherent noise level of the extra-high-gain signal tracer is comparatively high, as would be expected. However, as long as the signal output from the first mini-amp/spk is higher than its noise level, so will the sound output from the second mini-amp/spk be greater than the total noise level. In most cases, the volume control of the first mini-amp/spk will be advanced to maximum, and the volume control of the second mini-amp/spk will be advanced as required.

Figure 1-1

A useful audio-current checker/tracer. (a) application; (b) quick checker is more informative when provided with a calibrated voltage-gain scale; (c) isolating probe for use with mini-amp/spk; (d) extra-high gain audio signal-tracer arrangement for low-level applications.

"BALLPARK" TEMPERATURES[1]

An informative quick check of an audio amplifier is provided by temperature measurements of devices and components with a temperature probe or with a differential temperature checker. This quick check is most useful if you have a similar amplifier in normal working condition, for comparison purposes. However, even if you do not have a comparison unit available, much useful test data can be obtained.

For example, observe the typical normal temperatures for a driver transistor, and for a small power-type transistor, shown in Figure 1-2(a). With an ambient temperature of 18°C, the body of the driver transistor operates at 27°C; in a stereo amplifier, the driver transistors in the L and R channels will normally operate within 2° or 3° of each other—a substantial difference in temperatures points to a circuit defect.

Power transistors normally run hotter than low-level transistors; for example, the power transistor exemplified in Figure 1-2(b) operates at a heat-

---

[1]Caution: Power-type transistors can run hot enough to burn your finger.

Figure 1-2

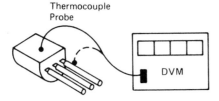

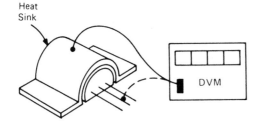

Pigtail typically runs 1°C hotter than the transistor body.

EXAMPLE: Ambient = 18°C
         Pigtail = 28°C
         Body = 27°C

(a)

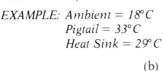

Pigtail typically runs 4°C hotter than the heat sink on the transistor body.

CENTIGRADE-FAHRENHEIT
CONVERSION SCALES

EXAMPLE: Ambient = 18°C
         Pigtail = 33°C
         Heat Sink = 29°C

(b)

Note: The hottest spot in a transistor is inside at its collector junction. On the exterior of the transistor, a temperature probe will show that the pigtails run hotter than the body of the transistor.

When a heat sink is used with a transistor, overheating can result from poor thermal contact of the transistor to the heat sink, or poor thermal contact of the heat sink to the chassis. Silicone grease may be used to ensure good flow of heat between contact surfaces.

Technical Note: "Jury riggers" sometimes force increased output from a transistor by using higher V_{cc} voltage than rated. This is very poor practice, and early failure of the transistor can be expected.

Figure 1-2 (continued)

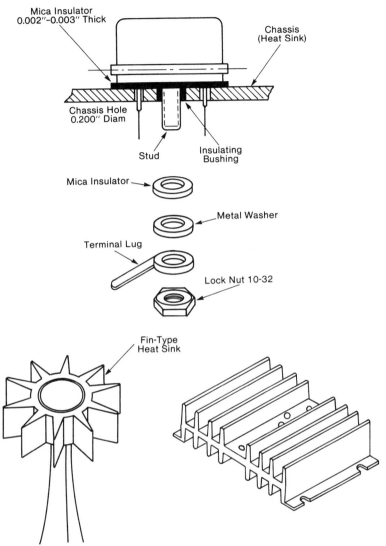

Mica Insulator 0.002″-0.003″ Thick

Chassis (Heat Sink)

Chassis Hole 0.200″ Diam

Stud

Insulating Bushing

Mica Insulator

Metal Washer

Terminal Lug

Lock Nut 10-32

Fin-Type Heat Sink

Reproduced by special permission of Reston Publishing Co. and Michael Thomason from Handbook of Solid-State Devices.

Figure 1-2

Example of transistor normal operating temperatures: (a) driver transistor; (b) small power transistor; (c) typical heat-sink arrangements. (Reproduced by special permission of Reston Publishing Company and Michael Thomason from *Handbook of Solid-State Devices.***)**

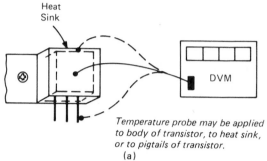

Heat Sink

DVM

Temperature probe may be applied to body of transistor, to heat sink, or to pigtails of transistor.

(a)

Example: In normal operation, with no signal input, and with an ambient temperature of 20°C, the body temperature of the transistor is 28°, the heat sink temperature is 28°C, and the pigtail temperature is 33°C. Again, in normal operation, with signal input for 2.5 watts output, the body temperature is 44°C, the heat-sink temperature is 44°C, and pigtail temperature is 46°C.

Note: Power output is equal to E^2/R, where E is the rms voltage of a sine-wave signal across the load, and R is the value of the load resistor. For example, 4.47 V rms across an 8-ohm load is equal to 2.5 watts.

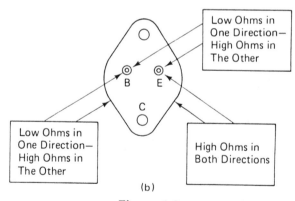

Low Ohms in One Direction— High Ohms in The Other

Low Ohms in One Direction— High Ohms in The Other

High Ohms in Both Directions

B   E

C

(b)

Figure 1-3

Output transistor quick checks: (a) example of output transistor operating temperatures in a 2.5-watt audio channel; (b) ohmmeter check of power transistor.

sink temperature of 29°C. Output transistors in the L and R channels of a stereo amplifier will normally operate within 3 or 4° of each other—a substantial difference in temperatures points to a circuit defect.

> **Subnormal operating temperatures indicate circuit malfunction, just as abnormal operating temperatures indicate circuit malfunction.**

Note that the normal operating temperature of an output transistor (in particular) depends significantly on whether it is tested *with signal*, or *without signal*. In Figure 1-3, observe that with no signal input, the heat-sink temperature of the output transistor is 28°C, whereas with signal input for a rated output of 2.5 watts, the heat-sink temperature is normally 44°C.

Another practical example is shown in Figure 1-4. Here, a power-type integrated circuit is the output device. When checked *without signal*, the heat-sink temperature is normally 25°C. On the other hand, when checked *with signal* for a rated output of 10 watts, the heat sink normally operates at 61°C, or over 140°F. (See Chart 1-1).

Figure 1-4

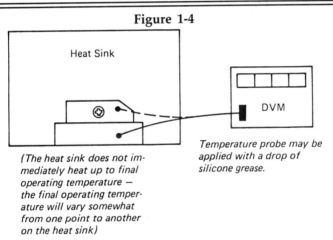

(*The heat sink does not immediately heat up to final operating temperature — the final operating temperature will vary somewhat from one point to another on the heat sink*)

Temperature probe may be applied with a drop of silicone grease.

Example: An integrated circuit with a rated power output of 10 watts normally idles (no signal input) at an ambient temperature of 19°C with an IC body temperature of 24°C, and a heat-sink temperature of 25°C. When operated with signal input for an output of 10 watts, the IC body temperature normally increases to 62°C, and the heat-sink temperature increases to 61°C (typical).

Although comparative temperature measurements are most informative when *identical* ICs are checked on a temperature probe, useful data can also be obtained by making comparative temperature measurements on nonidentical ICs that are rated for the *same power output.*

Figure 1-4 (continued)

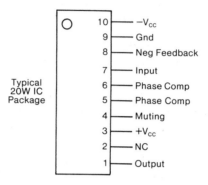

Note: IC power amplifier packages range up to 20 watts and are provided in single or dual versions. Most of them have an external heat sink to dissipate the relatively large heat rise. The majority are operated in class B. Many power amplifiers have automatic shutdown circuits to protect the IC when its rated temperature limit is reached. Short-circuit and overvoltage protection are also often included in an IC power amplifier.

Figure 1-4

Temperature check of power integrated circuit, and arrangement of a typical 20-watt IC package.

Chart 1-1

Integrated Circuits

Most modern electronic equipment is designed with specialized integrated circuits that provide functions of many discrete devices and components. When troubleshooting ICs without service data, it may be noted that twenty-four varieties of audio preamplifier packages are in general use.

Most of the IC packages that will be encountered are flat packs. However, the SK3166/949 8-pin TO-5 (round) package is also used in some amplifiers. Observe that flat packs may have from 7 to 16 pins. Two general types of flat packs are used, as exemplified below.

Chart 1-1 (continued)

 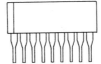

In various situations, the type number of the integrated circuit will be printed on the package. This is helpful information in preliminary troubleshooting procedures, inasmuch as the technician can refer to a standard IC data book for identification of each pin on the package. In turn, the preamp circuit can be rapidly "sized up," and DC voltage measurements can be made to "get a handle" on the problem.

In other cases, the type number of the IC will not be printed on the package, or the print may be illegible. Accordingly, a more basic approach to the preamp circuitry must be taken. Note that in the case of stereo preamps, two identical IC packages are often used (one in the R channel, the other in the L channel). In such a case, comparison tests with a DC voltmeter will frequently serve to pinpoint the fault, and to localize it either to the IC or to its external circuit components.

In the case of IC packages such as the SK3924/942, both the R preamp section and the L preamp section are contained in the same package. However, the same preliminary troubleshooting approach as described above applies for separate L and R packages.

Again, situations will be encountered in which the type number of the IC will not be printed on the package, or if printed, will be illegible. Further, trouble conditions may involve either "workable amplifier," or "dead amplifier." Basic approaches to these troubleshooting problems are discussed in Chapter 2.

NEW TEST EQUIPMENT FOR TROUBLESHOOTING WITHOUT SERVICE DATA

When troubleshooting electronic equipment without service data, various new kinds of test equipment may be used to speed up and simplify the job. For example, we will find that a DC or AC recording voltmeter can be easily constructed from a few components and used with an ordinary tape recorder to monitor voltage levels over any desired period of time. This new kind of test instrument is particularly helpful when troubleshooting "tough-dog" intermittent problems.

Again, we will find that an automatic internal resistance ohmmeter can be easily constructed to measure the "hot" resistance of any circuit from a chosen test point to ground. This new kind of test instrument can be very useful when "everything has been checked, but the trouble has not been found." Although dynamic internal resistance can be measured without an automatic ohmmeter, the traditional approach requires special connections, pairs of measurements, and calculation.

RESISTOR, CAPACITOR, DIODE, AND TRANSFORMER TEMPERATURES

In some trouble situations, highly informative preliminary checks can be made by measuring the temperatures of resistors, capacitors, diodes, and transformers. As exemplified in Figure 1-5, each component has its normal operating temperature. This temperature may be the same with signal or without signal, or, it may change from zero signal to full rated output signal level.

Figure 1-5

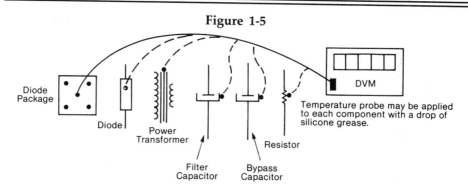

Diode Package

Diode

Power Transformer

Filter Capacitor

Bypass Capacitor

Resistor

DVM

Temperature probe may be applied to each component with a drop of silicone grease.

Example: A preliminary checkout of R, L, and C components and devices can be made with a temperature probe while the electronic unit is idling (no signal input). Although certain variations can be expected, these components normally operate at approximately the same temperatures in any unit of electronic equipment. (This is an example for a 20-watt PA amplifier.) With an ambient temperature of 20°C, "ballpark" temperature values for R, L, and C components and rectifier diodes are typically:

Power transformer, 24°C
Power-supply filter capacitor, 23°C
Electrolytic bypass capacitor, 22°C
Resistors, from 22° to 26°C
Rectifier diodes, 25°C

Figure 1-5 (continued)

Substantially higher temperatures point to an associated malfunction that causes excessive current drain.

Subnormally lower temperatures point to an associated malfunction that causes insufficient or zero current flow.

Note: The normal operating temperature of a diode depends on the amount of current that it is conducting. For example, a 1N34A germanium diode conducting 1 mA has a typical voltage drop of 277 mV and a temperature of 23°C (ambient temperature of 21°C). If the same diode is conducting 2 mA, its voltage drop increases to 317 mV, and its temperature increases to 24°C. Again, if the diode is conducting 3 mA, its voltage drop increases to 337 mV, and its temperature increases to approximately 25°C. Observe that if a diode is short-circuited or open-circuited, its temperature will rest at ambient.

Figure 1-5

"Ballpark" examples of normal operating temperatures for R, L, and C components, and for small diodes.

Additional helpful test data can be obtained if you are troubleshooting a stereo amplifier, because corresponding components in the L and R sections *normally* operate at practically the same temperature. Therefore, if a comparative temperature check shows that a given component in the L section is operating at a significantly higher or lower temperature than the corresponding component in the R section, the troubleshooter concludes that a malfunction is occurring.

Whether the malfunction is associated with the L component or with the R component is determined by these related facts:

1. If it has been established previously that the trouble symptom is localized to the L channel (for example), it is reasonable to conclude that the off-temperature component in the L channel is associated with circuit malfunction.

2. An "out-of-the-ballpark" temperature in itself is a definite indication of circuit malfunction—for example, a small resistor operating in the 60°C temperature range (instead of the 20°C range) points to circuit malfunction in either a stereo or mono amplifier.

3. When both the L and R channels exhibit trouble symptoms, and the temperatures of corresponding components are different (although "in the ballpark"), additional tests are required.

HIGH-SPEED DIFFERENTIAL TEMPERATURE CHECKER

The quick checker shown in Figure 1-6a shows whether a pair of corresponding devices are operating at the same temperature, or at different temperatures. For example, one diode may be placed on a reference IC, and the other diode may be placed on a corresponding suspected IC. If both ICs are operating at the same temperature, the DVM remains zeroed. On the other hand, if the ICs are operating at different temperatures, the DVM will indicate a positive voltage or a negative voltage.

The advantage of the differential temperature probe quick checker is its speed in application. (See Figure 1-6b.) The troubleshooter does not measure individual temperatures, but merely notes whether they are the same or not—that is, whether the DVM indicates zero, or a positive or negative reading.

Figure 1-6

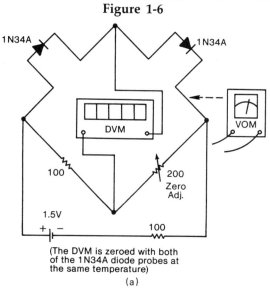

(The DVM is zeroed with both
of the 1N34A diode probes at
the same temperature)

(a)

Note: Observe that the battery is polarized so that both of the diodes conduct forward current in the temperature bridge. The forward resistance of a diode changes as its environmental temperature changes, or, the internal resistance of the diode decreases when its temperature is increased. In turn, the bridge becomes more or less unbalanced when the two diodes have different environmental temperatures. The value of the unbalance voltage is indicated by the DVM.

Quick checks can be facilitated by using a VOM instead of a DVM to indicate the bridge unbalance voltage. In other words, it is easier to observe the motion of a pointer on a scale than to consider the value of a readout on a DVM.

Figure 1-6 (continued)

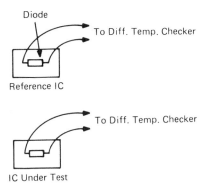

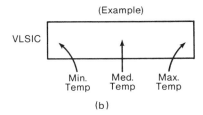

(b)

Note: When an IC with a comparatively large number of pins is under test, it is often important to place the diode probes at corresponding locations on the reference IC and the IC under test. In other words, a multistage IC may have significantly different temperatures at the input end, middle region, and output end. A preliminary check is made to best advantage by placing the diode probes in the middle regions of the reference IC and the IC under test.

Figure 1-6

Differential temperature quick checker: (a) bridge arrangement; (b) application of differential temperature probe.

NOTES ON INTEGRATED CIRCUIT PREAMPLIFIERS

We do not usually expect to find an integrated circuit used by itself in a preamplifier. Instead, an integrated circuit will ordinarily be preceded by a bipolar transistor. The reason for this combination is that an IC has an inherently high noise level, and a bipolar transistor has a comparatively low noise level, particularly when operated at a low collector voltage.

A typical preamp in this category uses an 8-pin IC and a small-signal input transistor. The arrangement provides a maximum voltage gain of approximately 1700 times, with 0.2 watt output from a 1-mV input. Most of the noise in an amplifier is contributed by the input stage, because noise from the input stage is amplified by all subsequent stages.

FOLLOW-UP DC VOLTAGE MEASUREMENTS

Follow-up DC voltage measurements are generally made to best advantage by buzzing out transistors in stages with malfunctions (or suspected malfunctions). As explained in Chart 1-1, the measured voltages may or may not make sense. It is often a question of whether nonsense voltages are being caused by a circuit defect or a transistor defect (or possibly both).[2]

The procedure generally is to disconnect the transistor from its circuit for test. (This is ordinarily done by unsoldering the transistor leads from the PC board, but it can also be accomplished by razor-cutting the PC conductors. After the test is completed, the cut conductors can be repaired with a small drop of solder.)

Troubleshooters may use a transistor tester to check out a transistor, or an ohmmeter may be used. Although the following tests are not really new, they are of fundamental importance, and should be carefully noted. With reference to Figure 1-7, a transistor is checked out-of-circuit with an ohmmeter as follows:

1. Measure the resistance between each pair of transistor terminals.
2. The two lowest resistance values are from base to emitter and from base to collector, thereby identifying the base terminal.
3. Whether the transistor is a PNP or an NPN type is shown by the polarity of the ohmmeter test leads in measurement of forward resistance.
4. Whether the transistor is a silicon or a germanium type is shown by the value of forward resistance, based on the troubleshooter's experience with the ohmmeter.
5. The collector and emitter terminals can be identified from the rule that a lower resistance is measured between these terminals when the test voltage is applied in normal operating polarity.

FINGER TEST

Unless the ohmmeter has megohm ranges, a finger test must be used to carry out Step 5 above. In other words, many ohmmeters cannot indicate the very high resistance between the collector and emitter terminals of a silicon

[2]See Chapter 10 for a discussion of progressive trouble-symptom development.

Figure 1-7

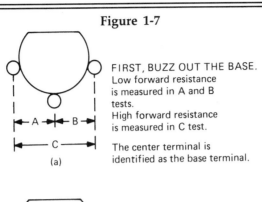

FIRST, BUZZ OUT THE BASE.
Low forward resistance
is measured in A and B
tests.
High forward resistance
is measured in C test.

The center terminal is
identified as the base terminal.

(a)

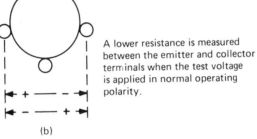

A lower resistance is measured
between the emitter and collector
terminals when the test voltage
is applied in normal operating
polarity.

(b)

Note: Transistor basing requires attention. When a transistor is replaced, the basing may be different from that of the original transistor, although their appearances are identical.

Case History: An NPN replacement transistor checked out with its collector and emitter terminals reversed, as compared with the original transistor. Moreover, *the basing diagram on the replacement transistor packet was incorrect.* Therefore, the troubleshooter should not assume that the basing of a transistor is the same as would be expected. *Always verify the basing with ohmmeter tests.*

Technical Note: The foregoing checkout procedure applies to conventional bipolar transistors. We will occasionally encounter Darlington transistors that look like conventional transistors. However, a Darlington transistor consists of a pair of "piggyback" transistors, with distinctive terminal resistance values, as explained subsequently.

Troubleshooters should also note that bipolar transistors may be unsymmetrical or symmetrical. In other words, an unsymmetrical transistor has an emitter forward resistance that is lower than the collector forward resistance. On the other hand, a symmetrical transistor has

Figure 1-7 (continued)

an emitter forward resistance that is equal to the collector forward resistance. An unsymmetrical transistor has much lower gain if its emitter and collector terminals are reversed. However, a symmetrical transistor has the same gain when its emitter and collector terminals are reversed. (An example of symmetrical transistor operation is shown in Figure 10-7.)

Reproduced by special permission of Reston Publishing Company and Walter Folger from Radio, TV, and Sound System Diagnosis and Repair.

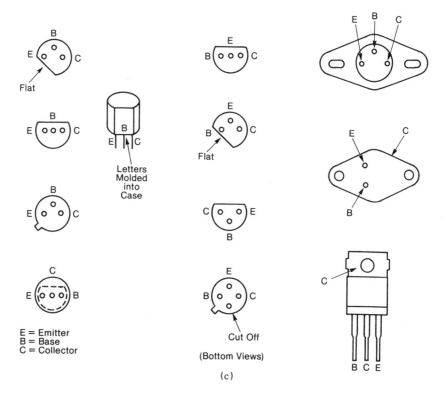

(c)

Technical Note: Some types of electronic equipment use phototransistors. From a practical point of view, a phototransistor is the same as a conventional transistor, except that it includes a lens on top of its case to admit light on its base-emitter junction. In most applications, only the collector and emitter terminals are used, and the base terminal is left "floating." An NPN phototransistor is operated with its collector positive and its emitter negative. If a phototransistor is con-

Figure 1-7 (continued)

nected to an ohmmeter in proper polarity, its collector-emitter resistance will fall to a very low value when light shines into the transistor.

Figure 1-7

Transistor checkout with an ohmmeter: (a) the base terminal has a low forward resistance to each of the two other terminals; (b) a lower resistance is measured between the emitter and collector terminals when the ohmmeter applies a voltage that is polarized as in normal operation (see also Finger Test); (c) these typical transistor basing arrangements show why an unidentified transistor should always be "buzzed out" with an ohmmeter to identify the base, emitter, and collector terminals. (Reproduced by special permission from Reston Publishing Company and Walter Folger from *Radio, TV, and Sound System Diagnosis and Repair*.)

transistor. However, a finger test may be used to provide resistance indication, no matter what kind of ohmmeter is used. To make a finger test, the troubleshooter proceeds as follows:

1. Apply the ohmmeter test leads to the collector and emitter terminals of the transistor (which is the collector and which is the emitter is unknown at this time).
2. Pinch the base lead and one of the other leads between the thumb and forefinger to provide "bleeder resistance." Note the resistance reading, if any.
3. Pinch the base lead and the remaining other lead between the thumb and forefinger to provide bleeder resistance. Note the resistance reading, if any.
4. Reverse the ohmmeter test leads and repeat Steps 2 and 3.
5. The collector is the terminal that provides the lowest resistance reading when its test voltage is bled into the base terminal.

Trick of the Trade: **If your skin is very dry, and you are using a 1000 ohms/volt meter, moisten your fingers slightly to bleed sufficient voltage into the base terminal.**

Caution: When checking circuit boards that have been loosened from the chassis, it is good foresight to cover chassis edges and protruding metal parts with masking tape. This insulation prevents accidental short-circuits to the solder side of the circuit board as it is moved about. The circuit board is usually connected to various front-panel controls and connectors, and to rear-chassis connectors with comparatively short leads. Accidental short-circuits are an occupational hazard.

Figure 1-8

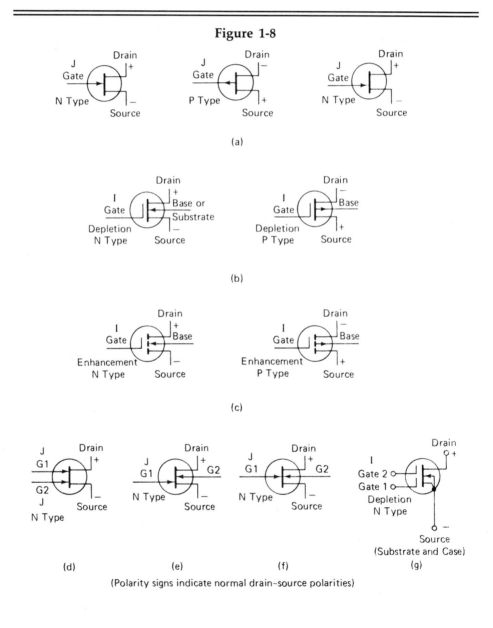

(Polarity signs indicate normal drain–source polarities)

Figure 1-8 (continued)

Field-Effect Transistors: (a) N and P types of JFETs—arrow points to N-type substrate, away from P-type substrate; (b) depletion type MOSFETs—arrow points to N-type substrate, away from P-type substrate; (c) enhancement-type MOSFETs—arrow points to N-type substrate, away from P-type substrate; (d) dual-gate N-channel FET; (e) nonsymmetrical N-channel FET; (f) alternate symbol for (e); (g) dual-gate depletion N-type MOSFET.

Caution: JFETs are comparatively rugged and can be handled and tested in much the same manner as bipolar transistors. On the other hand, MOSFETs are quickly damaged by static electricity whenever they are not in their circuits. Ground your wrist before you handle a MOSFET; ground the tip of your soldering iron before you solder a MOSFET into a circuit. Use heat sinks on both bipolar and unipolar transistor leads to avoid damage from overheating.

Reproduced by special permission of Reston Publishing Company and Michael Thomason from Handbook of Solid-State Devices.

Figure 1-8

Standard basing arrangements for field-effect transistors.

Case History: A momentary short-circuit between the chassis edge and the solder side of the circuit board blew the regulator transistor in the power supply, which had to be repaired before troubleshooting could continue.

FIELD-EFFECT TRANSISTORS

We occasionally encounter field-effect transistors (FETs) in audio equipment, and we need to know the various types in use. Field-effect transistors are also called *unipolar transistors* to distinguish them from the extensively used bipolar-type (junction-type) transistors.

If an FET is used in an audio unit, it will most likely be an insulated-gate FET (IGFET); an IGFET is also called a MOSFET, denoting its metal-oxide substrate field-effect construction. The various basing arrangements for common MOSFETs and IGFETs are shown in Figure 1-8. Note the handling precautions to avoid damage from static electricity or overheating.

It is advisable to keep a semiconductor reference guide at hand to quickly determine suitable replacements for defective semiconductors. Over 100,000 semiconductor substitutions are listed in the Archer (Radio Shack) Semiconductor Reference Guide No. 276-4006.

PARTS NUMBERING SEQUENCE

Amplifier circuit boards frequently have parts numbers marked beside devices and components. When troubleshooting without service data, it is often helpful to remember that in virtually all cases, part numbers start with "1" for input components and devices, and then increase sequentially through the network, with the highest part numbers marked beside output components and devices.[3]

IN-CIRCUIT TRANSISTOR TESTING

An in-circuit transistor tester, such as the B&K Precision 510 is often useful when troubleshooting without service data because it identifies all three transistor leads, shows whether the transistor is an NPN or PNP type, and whether it is a silicon or germanium type. The in-circuit tester provides good-bad indications with circuit shunt resistance as low as 10 ohms, and with shunt capacitance up to 15 μF. The instrument is easy to use because the test clips can be connected in any order to the transistor terminals. (See Figure 1-9.)

As a practical note, if a transistor tests "bad" in-circuit, it should be disconnected and then tested a second time out of circuit. This procedure ensures that a good transistor will not be discarded in the event that an unsuspected circuit fault might have been present during the in-circuit test.

FUNCTIONAL OVERVIEW OF CONSUMER-ELECTRONIC INTEGRATED CIRCUITS

A wide variety of integrated circuits (ICs) are encountered in electronic equipment, and it is helpful to recognize the basic functional types. Some ICs contain comparatively few transistors, resistors, and diodes. However, large-scale integrated circuits (LSIs) may contain tens of thousands of transistors in a package no larger than a postage stamp. We are chiefly concerned here with what ICs do and their circuit action in electronic equipment.

Note that integrated circuits are divided into four extensive categories. These are: *analog, consumer, digital,* and *interface.* From an applications viewpoint, these categories often overlap. As an illustration, a video cassette recorder (VCR) contains all four categories of integrated circuits. Subcategories may be considered, such as *amplifiers, comparators, converters, demodulators,* and so on. Further, subdivision may be made with respect to specific IC functions, such as *class A, differential,* or *operational amplifiers* to denote the level of application in electronic equipment.

[3]See Chapter 9 for a discussion of "buzzing out" circuits from the component side of the PC board.

Figure 1-9

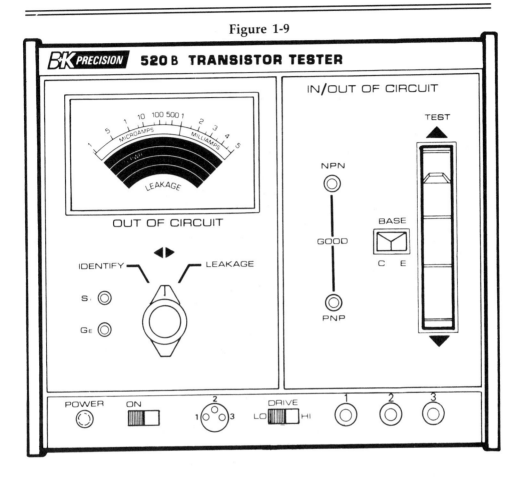

IN-AND-OUT-OF-CIRCUIT TRANSISTOR TESTER FUNCTIONS

1. Determines good or bad transistors, FETs, or SCRs in or out of circuit.

2. Determines good or bad diodes in or out of circuit.

3. Identifies emitter-base-collector leads of transistors.

4. Identifies gate lead of FETs.

5. Indicates polarity of good devices (NPN or PNP; N or P channel).

6. Identifies cathode-gate-anode leads of SCRs.

7. Indicates whether silicon or germanium device.

8. Measures I_{CES} or I_{BES} of transistors.

9. Measures I_{DSS} and gate leakage of FETs.

Figure 1-9 (continued)

10. Measures reverse leakage current of diodes.

11. Determines whether device is transistor, FET, or SCR.

Figure 1-9

A high-performance in-and-out-of-circuit transistor tester. (*Courtesy*, B&K Precision, Div. of Dynascan Corp.)

Analog ICs (also called linear ICs,) normally develop an output that is "like" the input signal (a similar or analogous output signal). In other words, a class-A amplifier is an analog arrangement, and an operational amplifier is an analog arrangement. On the other hand, although a class-A amplifier is always a linear unit as well as an analog unit, an operational amplifier (op amp) may not be a linear unit. Thus, if the op amp is used as a class-A audio amplifier, it is a linear unit and it is also an analog unit. However, if the op amp is used as an integrator, for example, it is then an analog unit only, and not a linear unit in the strict sense of the term. (The integral of the input voltage is a mathematical function of the input voltage, but it is not a linear function of the input voltage.) Digital ICs, of course, are all nonlinear devices.

At this point in our discussion, it is helpful to consider the basic parameters of a class-A amplifier IC:

1. *Gain.* Depending on the application and the type of amplifier under consideration, gain may be stated in terms of voltage gain, current gain, or power gain. Power gain may be stated as the voltage-current product, or it may be stated in decibels (dB). We will find that in various types of class-A amplifiers, gain may be controlled by variation of a DC control voltage. Or, manual gain control may be provided via a potentiometer connected to the gain line of the amplifier.

2. *Frequency (bandwidth).* Class-A IC amplifiers generally have a frequency response down to zero frequency (DC). However, the upper frequency limit, together with the maximum possible gain (MAG) at the rated upper frequency limit is often an important functional parameter. Note that the rated frequency limit for an IC does not take into account the limitations that might be imposed by external circuitry and which can greatly modify the rated MAG of the device itself.

3. *Linearity Distortion.* The rated linearity distortion for a class-A IC is generally stated as the maximum permissible input peak-to-peak signal voltage for the device. In various examples, the percentage of output signal distortion under specified operating conditions is also rated by the IC manufacturer. This is usually a total harmonic distortion (THD) value.

4. *Power Output.* Maximum rated power output for a class-A IC is given either in terms of dB with a specified load, or as the maximum peak-to-peak voltage across a specified load impedance. Beginners may note that a load impedance is associated with three forms of power values: The product of the rms AC voltage across the impedance and the rms AC current that flows through the impedance is called the *volt-ampere* power or the *apparent power*. It is stated in volt-ampere units. Again, the product of the apparent power and the power factor of the impedance is called the *real power* or the *resistive power*, or the *in-phase power*. Note that the power factor is equal to the cosine of the phase angle between the voltage applied to the impedance and the current flowing through the impedance. Finally, the product of the apparent power and the sine of the voltage-current phase angle is called the *imaginary power, reactive power,* or *quadrature power* in the impedance. *The maximum rated power output is stated with respect to the real power that is developed in the load impedance.*

5. *Noise.* The noise developed by an IC is rated as its signal-to-noise ratio (S/N) or as its noise figure. This rating denotes the amount of noise voltage that is generated within the IC when operating with a specified value of input resistance.

6. *Power Dissipation.* The maximum rated power dissipation for a class-A IC is stated in milliwatts or watts, and is related to the temperature rise of the IC package in normal operation. It is also related to heat-sink requirements. The value of the power dissipation is equal to the product of the DC voltage between V_{cc} and Gnd terminals and the DC current in the V_{cc} line. (This is the power dissipation under quiescent conditions, with no input signal applied.)

Observe the IC arrangement shown in Figure 1-10. This is a class-A amplifier arrangement. Pin numbers are indicated on the diagrams. We note in the equivalent circuit that two input terminals are provided. These input terminals are called the *inverting* input and the *noninverting* input. If the audio input signal is applied to the inverting input, the output signal will be 180° out of phase with respect to the input signal. On the other hand, if the audio input signal is applied to the noninverting input, the output signal will be in-phase with respect to the input signal.

The equivalent circuit also shows that both of the IC input terminals are above ground. In turn, both inputs may be driven by a push-pull (double-ended) signal, or, one input may be driven by a single-ended signal if the other input is returned to ground. The equivalent circuit further shows that there is a certain amount of input resistance, and that there is a certain amount of output resistance. Observe that a positive power supply (+V) and a negative power

Figure 1-10

Lag Compensation

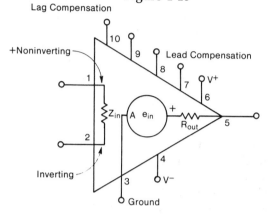

Equivalent Circuit

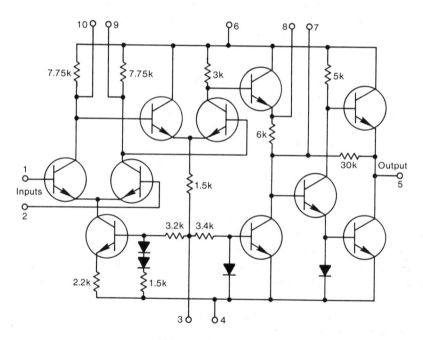

IC Internal Circuitry

Note: The IC comprises three functional sections, or stages. The input stage is termed the differential stage; the intermediate stage is termed the level-shifting stage; the final section is termed the output stage. Observe that the input stage is a differential configuration with a

Figure 1-10 (continued)

constant-current source in the emitter branch. (Optimum differential amplifier operation is obtained with constant-current bias.)

Figure 1-10

Arrangement of an IC used extensively in audio amplifiers.

supply ($-V$) are used with this IC. Both power supplies are returned to the ground terminal on the IC.

Note in the diagram for the internal circuitry of the IC that differential-amplifier stages are employed. (This design provides immunity to drift versus temperature changes.) The voltage gain of the IC per se is very high at zero frequency (DC) and then decreases with increasing frequency until the gain is very low at the upper end of the audio range. However, in application, a large amount of negative feedback is used (a resistor is connected from the output terminal to the inverting input terminal). This negative feedback reduces the gain, but greatly extends the frequency range of the amplifier and makes its frequency response essentially uniform (flat).

As the operating frequency is increased, stray capacitances inside of the IC cause phase shift of the output signal with respect to the input signal. Although phase shift is of no concern to the listener (phase distortion is commonly ignored in audio reproduction), phase shift becomes a problem in feedback circuit operation. In other words, if the phase shift becomes too great, the intended negative feedback becomes positive feedback, and the IC "takes off" and oscillates uncontrollably. For this reason, provisions are made for phase compensation. As indicated in the equivalent circuit, pins are provided for connection of a lag capacitor and for connection of a lead capacitor.

It is helpful to note that various *special-purpose* integrated circuits are used in consumer-electronics equipment. A special-purpose IC is essentially a subsystem that functions to replace several discrete-component stages in specific types of applications. (A discrete-component stage is arranged from individual transistors, diodes, and resistors.) Accordingly, a special-purpose IC provides multiple functions. Some typical special-purpose ICs are:

Television and FM IF amplifiers

Automatic fine-tuning systems

Audio amplifiers

Voltage regulators

Photodetector-and-power-amplifier circuits

Power-control circuits

An FM IF amplifier IC typically contains three cascaded differential-amplifier stages with a built-in voltage regulator. It has been manufactured in a 10-pin TO-5 style IC package. The IC is operated with external components and units comprising an FM tuner, tuned IF input transformer, various fixed capacitors, and an FM detector followed by an audio amplifier and speaker.

Next, an automatic fine-tuning system IC basically comprises four functional blocks: limiter-amplifier, balanced detector, differential amplifier, and regulator. An AGC amplifier block is included in a somewhat elaborated design. The balanced detector consists of four diodes. This IC is operated with external components and units comprising a TV tuner, tuned phase-detector transformer, tuned IF trap, and various fixed capacitors and resistors.

Note that a special-purpose audio amplifier IC typically consists of a 16-pin DIP containing four identical and independent amplifiers that can be connected, if desired, for operation in a dual-channel preamplifier for a hi-fi phono system. This IC is operated with external components comprising various resistors, fixed capacitors, and potentiometers (tone, level, and balance controls). In this application, the L preamp employs two of the amplifiers, and the R preamp employs the remaining two amplifiers in the IC.

Many special-purpose integrated circuits, other than those noted above, are encountered in modern electronic equipment. The more common types of special-purpose ICs are described in the chapters following. Descriptions are functional, inasmuch as the troubleshooter is primarily concerned with the circuit action and purpose of a given integrated circuit. This knowledge is almost indispensable when troubleshooting without service data.

OSCILLOSCOPE CAPABILITIES

In various "tough-dog" situations, time can often be saved by using a good oscilloscope to supplement the test data obtained otherwise. An oscilloscope is a form of voltmeter that displays a varying voltage as a function of time. In turn, it provides measurement of frequency, phase, rise time, interference, waveform period, pulse width, duty cycle, repetition rate, peak voltage, peak-to-peak voltage, average voltage, percentage modulation, damping time, and input/output relations.

Oscilloscopes are basically classified as free-running or triggered-sweep types. An oscilloscope with a free-running time base is adequate for considerable troubleshooting requirements in analog circuitry. On the other hand, oscilloscopes with triggered-sweep time bases and calibrated sweeps are a necessity in digital circuit troubleshooting procedures, for example. The beginner will find it helpful to note the following descriptive overview of a typical wide-band triggered-sweep oscilloscope. (See also Figure 1-11.)

It is good practice to turn the power off from a PC board before an oscilloscope is connected to any test point (particularly to an integrated-circuit pin). The reason is that it is almost impossible to avoid making an accidental short-circuit on occasion; if the power is turned on, the result is likely to be a burned-out IC.

A good knowledge of "how to read oscilloscope waveforms" is most helpful when troubleshooting electronic equipment without service data. Every distorted waveform has characteristics that point to the nature of the malfunction, and that may pinpoint the faulty component or device. Note in passing that most waveforms in electronic circuitry have a DC component in addition to the AC waveform. If the troubleshooting switches the scope from its AC function to its DC function, the vertical shift of the wavform shows the DC voltage that is present.

Figure 1-11

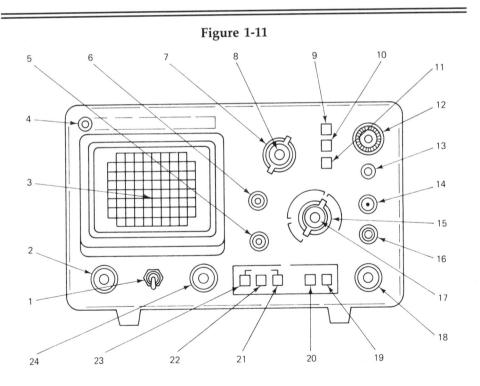

Figure 1-11 (continued)

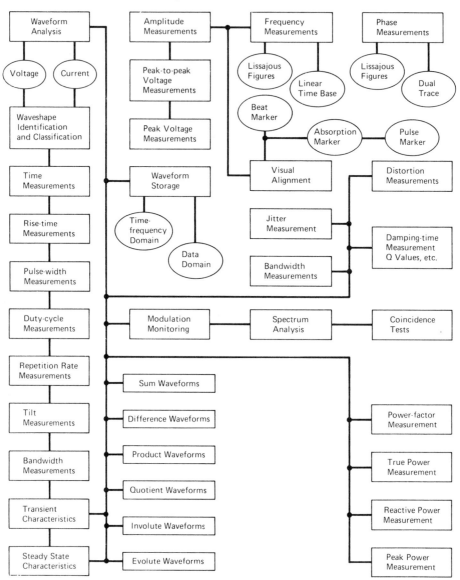

Figure 1-11

A modern solid-state triggered-sweep oscilloscope.
(*Courtesy*, B&K Precision, Div. of Dynascan Corp.)

2

ADDITIONAL AUDIO TROUBLE-SHOOTING TECHNIQUES

Audio Signal Tracers • Two-Tone Audio Signal Tracer • Recording Signal Tracer • Extended High Audio-Frequency Signal Tracer • Extended Low Audio-Frequency Signal Tracer • Signal Tracer with Recording Voltmeter • DC Voltage Monitor with Audible Indication • Recording DC Voltage Monitor • Emitter Follower for DC Voltage Monitor • Recording AC Voltage Monitor • Basic Amplifier Classifications • Troubleshooting Low-Power Audio Integrated Circuits • IC Functional Overview • Transistor Testing with the Oscilloscope

AUDIO SIGNAL TRACERS

Troubleshooting electronic equipment without service data is greatly facilitated by the availability of specialized types of signal tracers. These testers range from very simple (but highly practical) arrangements to sophisticated instruments that provide both qualitative and quantitative information, supplemented by signal-processing techniques. (See Table 2-1.)

Although elaborate audio analyzers are manufactured for laboratory applications, very few audio signal tracers or general-purpose signal tracers are available commercially for the service technician, hobbyist, and experimenter. Accordingly, nonprofessional electronics workers are usually forced to "roll their own." Fortunately, build-it-yourself signal tracing equipment is not unduly difficult to construct. The simplest types of signal tracers, such as those shown in Figure 2-1, require no work other than provision of a pair of test leads.

As noted in Table 2-1, the basic amplifier-speaker unit may be supplemented with a dB meter (VOM) as indicated in Figure 2-1. This arrangement can be helpful in comparing signal levels from one test point with another. Note that the speaker will be disabled when the dB meter is plugged into the mini-amp. In other words, a relative sound level cannot be measured while the speaker is operating, and vice versa.

Although most VOMs have a dB scale for use on the AC-voltage ranges, they do not necessarily have sufficiently low AC-voltage ranges for this application. Stated another way, if the lowest AC-voltage range is 10 V rms, less than half-scale deflection will be obtained when used with the indicated mini-amp. Accordingly, a VOM that has a 2.5 or 3 V rms range should be used.

As noted in Figure 2-1, a Darlington emitter follower unit can be used ahead of the mini-amp, if desired, to obtain much higher input impedance. It is preferable to construct the emitter follower in probe form, so that the input lead to the transistor is comparatively short. In other words, this is a high-impedance lead; associated capacitance should be minimized to avoid high-frequency attenuation. In particular, coaxial cable should not be used; its capacitance could objectionably load a high-impedance audio circuit.

If very little distortion is present, the volume control of the amplifier should be turned up as required to make the residue signal clearly audible.

As a practical construction consideration, the selectivity (also called the discrimination) of the parallel-T RC filter depends greatly on R and C values. Unless very tight tolerance resistors and capacitors are used, the selectivity may be disappointing.

Table 2-1

Audio Signal Tracers

Qualitative Testers

* Basic amplifier/speaker units, such as depicted in Figure 2-1.

* "Timbre" tester, such as shown in Figure 2-2.

* "Timbre" tester with residue analyzer, as exemplified in Figure 2-3.

* Two-tone signal tracer for tracking down the source of frequency distortion (see Figure 2-4).

* Recording signal tracer (basic signal tracer and tape recorder for monitoring intermittent amplifiers over an extended period of time).

* Basic signal tracer with Darlington probe for checking in unusually high-impedance circuitry.

* Extended audio-frequency signal tracer (includes heterodyne action to make highest audio frequencies readily audible).

* Extended low audio-frequency signal tracer (includes modulator action to make lowest audio frequencies readily audible).

* Signal tracer with built-in linearity checker to indicate whether circuit under test is producing serious distortion.

* Programmable signal tracer with facilities for making sequential tests of the signal characteristics.

Quantitative Testers

* Basic amplifier/speaker units supplemented by a dB meter, such as shown in Figure 2-1.

* Calibrated oscilloscope.

* Basic signal tracer supplemented by an audio recording voltmeter.

* Signal tracer with built-in harmonic distortion meter to indicate percentage of distortion at the point of test.

Note: Recording instruments are of particular utility when troubleshooting tough-dog intermittent equipment. The recording facility enables equipment monitoring over any desired period of time unattended by the technician. In turn, the sporadic output(s), if any, can be evaluated at any future time.

It is helpful to keep the common causes of intermittents in mind:

1. Cold solder joints
2. Broken, cracked, or charred circuit boards
3. Defective plated-through holes or eyelets
4. Corrosion or metal fatigue in solder lug or lock-washer connections
5. Internal intermittents in transistors or integrated circuits
6. Internal intermittents in capacitors or resistors

Figure 2-1

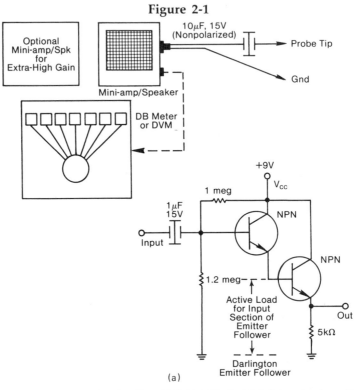

(a)

Note: The Micronta (Radio Shack) LCD Digital Multimeter is suitable for this application.

Technical Note: If different types of transistors are used, the base bias resistors may need to be changed in value. Archer (Radio Shack) No. 276-1617 was used in this configuration.

Note: The mini-amp/spk may be an Archer (Radio Shack) No. 277-1008B. It provides a gain of approximately 1700 times. A 1-mV input signal produces a 200 mW output.

(Most VOMs have a dB scale for use on the AC voltage ranges.) The dB meter is plugged into the earphone jack on the mini-amp housing. Observe that an input impedance of 5 kilohms will not objectionably load the majority of audio circuitry that we encounter. However, if excessive loading happens to occur in MOSFET circuitry, for example, the emitter follower shown in the diagram can be placed ahead of the signal tracer, and the input impedance thereby stepped up to 0.5 megohm.

Figure 2-1 (continued)

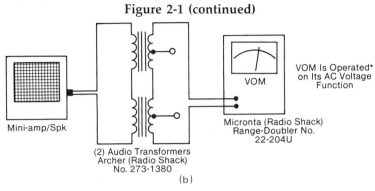

VOM Is Operated*
on Its AC Voltage
Function

VOM

Mini-amp/Spk

Micronta (Radio Shack)
Range-Doubler No.
22-204U

(2) Audio Transformers
Archer (Radio Shack)
No. 273-1380

(b)

Note: This method for increasing the sensitivity of a dB meter employs a pair of miniature audio transformers for impedance-matching and for stepping up the output voltage from the mini-amp/spk. In this example, the VOM is operated on its V-A/2 function. Observe that when the transformers are not used, the mini-amp will not provide full-scale deflection of the VOM, even if the mini-amp is operated at maximum output. On the other hand, with the transformers connected into the circuit, full-scale deflection of the VOM is obtained even though the mini-amp is operated at comparatively low output.

The mini-amp is designed to operate with a 16-ohm output load, and each of the transformers has an 8-ohm input impedance (if the secondary is loaded by a 1-kilohm impedance). Thus, with two primaries connected in series-aiding, their total input impedance is 16 ohms (if the series-aiding secondaries are loaded by a 2-kilohm impedance). The input impedance of the VOM in this example is 2.5 kilohms/volt. Accordingly, the impedance match is only approximate. However, the improvement in sensitivity is adequate for the application.

If you are using some other type of VOM in this arrangement, and the sensitivity of dB response is unsatisfactory, the practical procedure is to experiment with various types of audio transformers for determination of the optimum turns ratio.

Figure 2-1

**Audio signal tracer with level monitor: (a)
connection arrangement and optional high-input
impedance Darlington configuration; (b) practical
method for increasing the sensitivity of a dB
meter.**

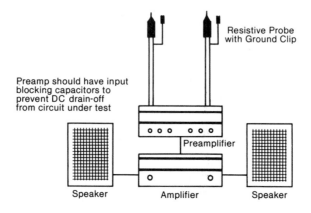

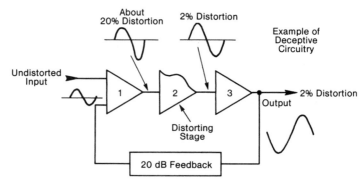

Feedback loop makes the first stage "look like" the distorting stage.

Reproduced by special permission of Reston Publishing Company, from
Handbook of Solid-State Troubleshooting, *by Hershal Gardner.*

Figure 2-2

**A "timbre tester" signal tracer consists of a mono
or stereo hi-fi arrangement with probe(s).**

Figure 2-3

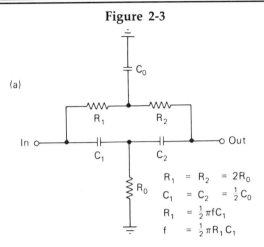

(a)

$$R_1 = R_2 = 2R_0$$
$$C_1 = C_2 = \tfrac{1}{2}C_0$$
$$R_1 = \tfrac{1}{2}\pi f C_1$$
$$f = \tfrac{1}{2}\pi R_1 C_1$$

Note: This is an effective audio distortion tracer and quick checker that employs a parallel-T RC filter (trap). In application, a 1-kHz test signal is used, and the switch is first set to apply the voltage from the test point directly to the hi-fi amplifier and speaker. This provides the reference tone from the speaker. Then the switch is thrown to place the RC network in series with the voltage from the test point and the hi-fi amplifier and speaker. This provides the comparison tone (if any) from the speaker. In other words, if there is no distortion in the equipment under test, the RC network will trap out practically all of the 1-kHz voltage and there will be little or no sound output from the speaker. On the other hand, if there is distortion present, there will be more or less sound output from the speaker for the comparison tone. This comparison tone will lack the 1-kHz fundamental frequency component, but will contain all of the distortion products, such as a second harmonic, a third harmonic, and so on.

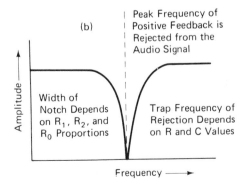

(b)

| Peak Frequency of
| Positive Feedback is
| Rejected from the
| Audio Signal

Amplitude

Width of
Notch Depends
on R_1, R_2, and
R_0 Proportions

Trap Frequency of
Rejection Depends
on R and C Values

Frequency ⟶

Figure 2-3 (continued)

When high-impedance audio circuitry is under test, the comparatively low input impedance of the parallel-T RC filter may load the circuit objectionably. In such a case, use the emitter-follower arrangement shown in Figure 2-1 ahead of the filter.

Note: A parallel-T RC filter can be arranged to reject any desired frequency. In the standard (basic) configuration, $R1 = R2 = 2R_o$, $C1 = C2 = C_o/2$, $R1 = (2\pi fC1)$, and the rejection frequency is $f_{rej} = 1/(2\pi R1C1)$. It is advisable to first assign a value for C1, and to calculate the corresponding value for R1. In turn, the value of f_{rej} can be calculated. Or, if R1C1 is calculated from a specified value of f_{rej}, the corresponding R1/C1 ratio can be calculated. In either case, we start with values for R1 and C1, which are also the values for R2 and C2. In turn, the values for R_o and C_o can be calculated.

The practical advantage of a parallel-T RC filter in troubleshooting procedures is the elimination of inductors from the filter circuitry. Inductors are undesirable, particularly at lower frequencies, because of their bulk and susceptibility to stray-field pickup.

Reproduced by special permission of Reston Publishing Co. and Walter Folger from Radio, TV, and Sound System Diagnosis and Repair.

Figure 2-3

Audio distortion tracer and quick checker. (a) component relations; (b) frequency response.

Figure 2-4

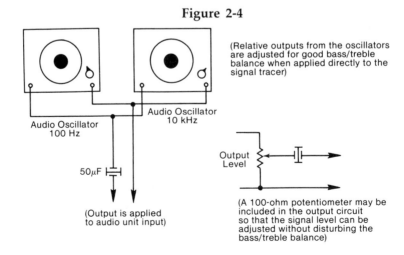

(Relative outputs from the oscillators are adjusted for good bass/treble balance when applied directly to the signal tracer)

Audio Oscillator 100 Hz

Audio Oscillator 10 kHz

$50\mu F$

Output Level

(Output is applied to audio unit input)

(A 100-ohm potentiometer may be included in the output circuit so that the signal level can be adjusted without disturbing the bass/treble balance)

Figure 2-4 (continued)

Note: When this two-tone signal source is provided with a signal tracer that has good response at 100 Hz and at 10 kHz, the arrangement operates as a useful frequency-distortion quick checker to track down the source of frequency distortion in an audio system or network. The troubleshooter "plays it by ear," and evaluates the output from the signal tracer for bass/treble balance.

Most troubleshooters can hear a 10-kHz tone, although some may have difficulty. If necessary, the higher test frequency may be reduced to 8 or 9 kHz. A more elaborate test arrangement may be used, if desired, to make very high test frequencies or very low test frequencies clearly audible, as explained subsequently.

Figure 2-4
Two-tone audio signal source for use with a two-tone signal tracer.

One easy expedient to obtain high selectivity with low-tolerance components is to use potentiometers instead of 500-ohm and 1000-ohm resistors. In turn, the potentiometers can be adjusted as required for optimum selectivity. As a guideline, it may be observed that when the filter is carefully constructed, the fundamental of the test signal will be practically eliminated, the second harmonic will be attenuated about 8 dB, and the third harmonic will be attenuated about 6 dB. Although these harmonic residues are somewhat attenuated, the amplifier volume control can be advanced to compensate for the insertion loss of the RC filter.

TWO-TONE AUDIO SIGNAL TRACER

Audio troubleshooters are concerned with two principal types of distortion. Amplitude distortion is evaluated with a timbre tester. Frequency distortion is also apparent in a timbre test, although a more informative quick check for frequency distortion can often be made with a two-tone signal tracer. This arrangement comprises the tracer shown in Figure 2-2 with the signal source depicted in Figure 2-4. The effectiveness of a two-tone quick check is in the loss (often extreme loss) of bass/treble balance that is generally associated with circuit faults that cause frequency distortion.

Many technicians have difficulty in hearing signals with frequencies much higher than 10 kHz. If difficulty is encountered in evaluation of a 10-kHz test signal, an 8-kHz signal can be used instead, for example. This reduction will not greatly impair the usefulness of the bass/treble balance test. Quick checks in this category do not provide quantitative conclusions; they are intended only to answer the question "Is the distortion greater or less at this point than at the previous test point?"

RECORDING SIGNAL TRACER

Troubleshooting intermittent audio equipment is a time-consuming procedure at best. Accordingly, it is helpful to use testers that save as much time as possible. For example, if a preamp operates satisfactorily most of the time, but has a history of occasional intermittents, the technician may feel that he or she has replaced the intermittent component or device—but is not sure. In this situation, the repaired amplifier must be "cooked" over an extended period of time to ensure that the fault has indeed been corrected.

As an example, the technician may opt to "cook" the repaired amplifier overnight to determine whether intermittent operation may occur over this period. In turn, the recording signal tracer does yeoman service. If the signal stops at any time, this failure is recorded on the tape, which can be checked at any convenient time. The tape recorder is plugged into the earphone jack of a mini-amp signal tracer, in the same manner as the dB meter depicted in Figure 2-1.

If two or three recording signal tracers are applied at successive test points in the intermittent preamp, the trouble area can be narrowed down accordingly. As a practical consideration, an open-reel tape recorder with a large reel should be used if the monitor is to be left unattended for a considerable length of time. It is also helpful to operate the recorder on its slowest speed, in order to make the tape "last as long as possible."

Another basic type of signal-recording requirement occurs when a preamp is "dead most of the time, but comes on once in a while." In this situation, monitoring over an extended period of time is required, preferably with two or three recording signal tracers to narrow down the trouble area. (Each signal tracer is applied at a different test point.) As a practical consideration, a voice-actuated tape recorder should be used in this case, since there is only sporadic output from the intermittent preamp. The recording on each signal tracer will then start only when a signal voltage is present. (A Realistic Radio Shack CTR-75 recorder is suitable for this application).

EXTENDED HIGH AUDIO-FREQUENCY SIGNAL TRACER

As noted previously, many technicians cannot hear tones with a pitch much higher than 10 kHz. Very few people can hear a 15-kHz tone, and a very loud 20-kHz tone is sensed primarily as a "feeling of pressure." Since the majority of high-fidelity amplifiers are rated for response out to 20 kHz, troubleshooters need an extended audio signal tracer to track down high-frequency distortion in the range from 10 kHz to 20 kHz. This need can be met easily with a suitable two-tone signal tracer that has heterodyne action.

With reference to Figure 2-4, a two-tone signal is provided; in this application, one audio oscillator is set to 20 kHz, and the other audio oscillator is set to 19.5 kHz, for example. The difference frequency, 500 Hz, can then be made plainly audible with a suitable nonlinear probe used with a conventional signal tracer. (See Figure 2-5.) The relative loudness of the difference frequency from one test point to another clearly shows where extended high-frequency loss is occurring and guides the technician to the trouble area.

Note that a demodulator probe should not be used in ordinary signal-tracing procedures that employ single-tone tests. The chief disadvantage of a demodulator probe in conventional signal-tracing tests is the introduction of serious distortion. Moreover, a substantial insertion loss will occur. In other words, if single-tone tests are being made, either a direct probe or a resistive probe (or a Darlington emitter follower) should be utilized.

EXTENDED LOW AUDIO-FREQUENCY SIGNAL TRACER

Just as the ear is at a disadvantage in hearing extended high-frequency audio tones, so is it at a disadvantage in hearing extended low-frequency audio tones. Inasmuch as high-fidelity preamps are rated for response down to 20 Hz,

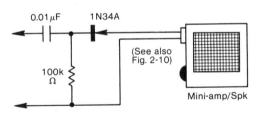

Note: This is a basic demodulator probe arrangement for use with a high-frequency audio signal tracer. Observe that although a 20-kHz signal and a 19.5-kHz test signal are both inaudible, their difference frequency, 500 Hz, is developed by the demodulator probe and becomes plainly audible. The relative level of the 500-Hz tone from one test point to another indicates the circuit response to a 20-kHz signal.

The 1N34A diode is a germanium type, and has an advantage in low-level tests because its barrier potential is approximately half of a silicon diode's barrier potential.

Figure 2-5

Basic demodulator probe configuration for use with a conventional signal tracer in extended high-frequency audio signal-tracing procedures.

the troubleshooter needs a specialized signal tracer for tracking signals in the 20 to 100-Hz range. That is, although the troubleshooter can hear a 20-Hz signal, for example, its intensity is comparatively low, and it is difficult to judge its change in intensity from one test point to another.

This need is easily met by means of a signal-tracing arrangement with modulator action, as depicted in Figure 2-6. Observe that the output signal from the 500-Hz audio oscillator is amplitude-modulated by the output from the 20-Hz audio oscillator. In turn, the troubleshooter evaluates the level of the 20-Hz signal at a test point in terms of the loudness of a rising and falling 500-Hz tone. The effectiveness of this method is the high receptivity of the ear to a 500-Hz tone, whereas the ear has very low receptivity to a 20-Hz tone. The tone-modulator arrangement is used with an ordinary signal tracer.

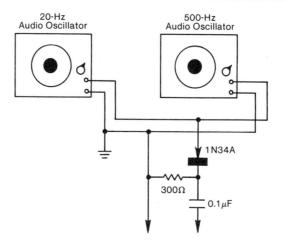

Note: The two audio oscillators are connected in parallel and their combined outputs are fed through a 1N34A germanium diode. In turn, the 500-Hz signal is amplitude-modulated by the 20-Hz signal. Thus, the intensity of the 500-Hz signal rises and falls at a 20-Hz rate. The outputs from the audio oscillators should be adjusted to about the same level, and overload of the diode should be avoided.

Figure 2-6

A simple and effective tone modulator for use with an extended low-frequency signal tracer.

SIGNAL TRACER WITH RECORDING VOLTMETER

Intermittent signal monitoring over comparatively long periods of time becomes more informative when pertinent data such as voltage levels are recorded. Either DC or AC voltage levels (or both) can be recorded when the signal tracer is combined with suitable arrangements. In turn, the troubleshooter may leave an intermittent preamp unattended for any period of time, and later evaluate the tape(s) for occurrences of an intermittent, along with voltage data that assist in pinpointing the fault.

DC VOLTAGE MONITOR WITH AUDIBLE INDICATION

A useful DC voltage monitor is illustrated in Figure 2-7. This tester frees the user from glancing repeatedly at a meter over an extended period of time while making circuit adjustments, or experimental replacement of components or devices. This monitor is particularly helpful when an intermittent unit must

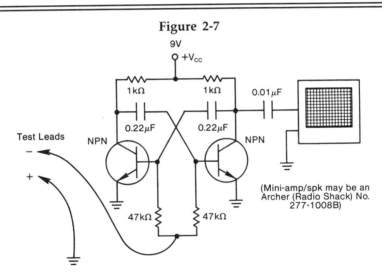

Figure 2-7

Note: The astable multivibrator operates as a voltage-controlled oscillator; its output frequency is reproduced as an audible tone by the mini-amp. The pitch of the tone varies over a wide range as the voltage applied to the test leads ranges from ground potential to −6 volts, for example. As the negative input voltage increases, the pitch of the tone increases. In turn, the arrangement serves as a practical voltage monitor that liberates the user from watching a meter over an extended period of time.

Figure 2-7 (continued)

Technical Note: The transistors used in this experimental arrangement were Archer (Radio Shack) No. 276-1617. However, any general-purpose NPN transistors may be used, provided that the circuit resistances are suitable. In particular, the base resistance value may need to be changed from 47 kilohms to obtain a suitable input/output response range.

Figure 2-7

DC voltage monitor (VCO) with audible indication.

be monitored over an indefinite interval, awaiting onset of the intermittent malfunction. Thus, the operator can connect the monitor at a key point in the intermittent unit, and then go about another project. Later, when the intermittent condition occurs, the operator is alerted immediately by the change in pitch of the sound output from the monitor.

This particular configuration is intended for monitoring negative voltages. Thus, a −1 volt potential produces a higher pitch in the sound output than a zero-volt potential does. Again, a −2 volt potential produces a higher pitch in the sound output than a −1 volt potential does, and so on. Positive voltages may be monitored by a slightly different configuration, which will be explained subsequently. Observe that the input impedance to this basic monitor is comparatively low, and may tend to load high-impedance circuitry. The effective input impedance to the monitor can be increased greatly with the addition of an emitter follower, as will be explained.

RECORDING DC VOLTAGE MONITOR

The monitor arrangement shown in Figure 2-7 may be operated as a recording voltmeter when it is supplemented by a tape recorder. That is, a tape recorder may be placed near the monitor while a voltage level is being checked over an extended interval. This is a particularly useful procedure when an intermittent unit is being checked, for example. Stated another way, the operator does not need to remain within hearing distance of the monitor because the recorded tone track can be played back for analysis at any later time.

Silent recording may be employed, if desired, by connecting a patch cord between the output of the mini-amp and the input of the tape recorder. Observe also, that if a stereo tape recorder is utilized, two DC voltages can be monitored simultaneously and recorded on the L and R tracks. This is a particularly helpful procedure when a comparison test is to be made of a malfunctioning unit and a similar comparison unit in normal operating condition. Then, during playback at some later time, any discrepancy in pitch between the two monitor outputs becomes clearly evident.

EMITTER FOLLOWER FOR DC VOLTAGE MONITOR

As noted previously the input resistance to the basic DC voltage monitor is comparatively low. Accordingly, when checking high-impedance circuitry, it may become desirable to increase the effective input resistance of the monitor with an emitter follower as shown in Figure 2-8. This arrangement is suitable for monitoring negative DC voltage levels. On the other hand, to monitor positive DC voltage levels, the emitter-follower configuration depicted in Figure 2-9(a) should be employed. Observe that the ground bus in the emitter follower will be at the same potential as the ground bus in the unit under test. However, the ground bus in the DC voltage monitor will be at the emitter potential of the lower transistor in the emitter follower.

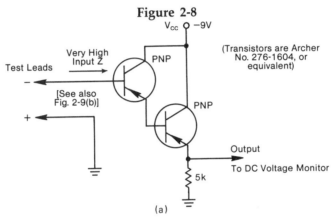

Figure 2-8

(a)

Note: The basic distinction between this arrangement and the configuration shown in Fig. 2-1(a) is that the former is a-c coupled, whereas this arrangement is direct-coupled.

Technical Note: An emitter follower operates as a current amplifier, and its output voltage is somewhat less than its input voltage. This is an example of the Darlington connection in an emitter follower, which provides very high current gain. In turn, the input impedance to the emitter follower is very high. Thus, when the DC voltage monitor is preceded by this emitter-follower section, tests can be made in very high impedance circuits without objectionable loading. Observe that this arrangement is suitable for monitoring negative DC voltages.

The Darlington arrangement is also called a beta multiplier because its current gain is roughly equal to the product of the individual transistor current gains. Thus, if the DC beta of each transistor is 100, then the beta of the Darlington circuit will be approximately 10,000.

Figure 2-8 (continued)

Observe that the output impedance (resistance) of the Darlington circuit is quite low (considerably less than the value of the emitter load resistor) due to inherent negative-feedback circuit action.

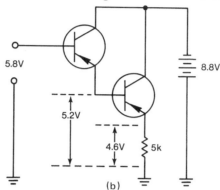

(b)

Note: Although an emitter follower is a current amplifier, it is a voltage "divider." In other words, the output voltage is less than the input voltage. Thus, in this example, the input voltage is 5.8 V, which is reduced to 5.2 V at the output of the first transistor. In turn, the 5.2 V at the input of the second transistor is reduced to 4.6 V at its output (across the emitter load resistor). Observe that this voltage division is independent of the value of the collector supply voltage. The V_{cc} value merely determines the upper limit of input voltage that can be applied for normal circuit operation.

Technical Note: To see why the input impedance of the emitter follower is very high, observe the following circuit relations:

1. The input voltage is applied between the base of the upper transistor and ground.
2. The base draws current, and this input current is multiplied by the beta factor of the circuit.
3. The multiplied input current produces a comparatively large output voltage that cancels most of the input voltage.
4. In turn, only a very small input current can flow, and the input impedance (resistance) V_{in}/I_{in} becomes very high.

Figure 2-8

Basic direct-coupled Darlington emitter follower: (a) very high input impedance and very low output impedance are provided; (b) example of DC voltage distribution in a Darlington emitter-follower circuit.

Figure 2-9

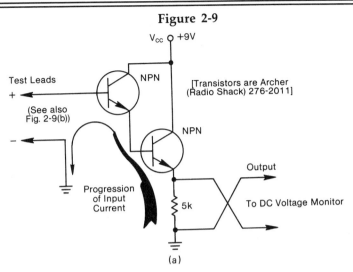

(a)

Note: This configuration is similar to the arrangement shown in Figure 2-8 except that NPN transistors are utilized. In turn, this emitter follower provides for monitoring positive DC voltage levels. It provides very high input impedance as a result of the Darlington-connected transistors. Observe that the emitter of the lower transistor becomes more positive as the input is driven more positive. Accordingly, the ground lead of the emitter follower becomes more negative as the input is driven more positive. The practical significance of this relationship is that this emitter follower must utilize a separate collector supply battery and that its ground connects to the 47k resistors in Figure 2-7. Consequently, the ground in this DC voltage monitor

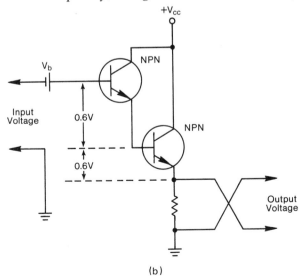

(b)

Figure 2-9 (continued)

must "float" (a common ground must not be used between this emitter follower and the DC voltage monitor).

Note: An emitter follower of any kind does not output any voltage until the base-emitter junctions are forward biased. Accordingly, there is a lower limit of input voltage that can be monitored by the elementary Darlington configuration. When we need to monitor a voltage level less than 1 volt, for example, it is necessary to include a bias battery V_b in the input lead, as shown above. If V_b has a value of 1.5 volts, 1.2 volts will then be available for forward bias with zero input and the test leads short-circuited (or connected into a conductive circuit). Observe that this principle applies also to the arrangement in Figure 2-8 (with the polarity of V_b reversed).

Figure 2-9

Darlington emitter-follower arrangement for monitoring positive DC voltage levels: (a) circuit; (b) biased configuration permits monitoring of low DC voltage levels.

RECORDING AC VOLTAGE MONITOR

When an intermittent condition occurs in an audio preamp, the signal may not disappear entirely; instead it may drop to some subnormal level, such as one-half, one-tenth, or one-fiftieth of its normal value. Knowledge of the amount of attenuation that occurs during the intermittence can provide useful clues in pinpointing the circuit fault. An AC voltage monitor is very similar to a DC voltage monitor, except that it contains one additional feature: a rectifier probe such as the one depicted in Figure 2-10.

Figure 2-10

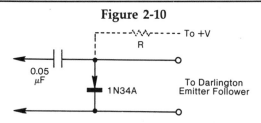

Note: This is a basic rectifier probe arrangement that develops the negative peak voltage of an audio signal. In turn, the output from the probe can be fed into the emitter-follower configuration shown in Figure 2-8. Thus, the DC voltage monitor is converted into an AC voltage monitor. The unbiased probe is useful chiefly at signal-voltage levels above 0.25 volt, due to the barrier potential of the germanium diode.

Figure 2-10 (continued)

Technical Note: If low-level AC voltages are to be monitored, the sensitivity of the basic probe can be improved by means of a bias resistor as shown by the dotted lines in the diagram. Bias resistor R has a high value and is connected to any convenient source of positive DC voltage. The value of R is experimentally chosen for optimum probe sensitivity to low-level AC signals.

Figure 2-10

A rectifier probe suitable for use with an AC voltage recording monitor.

The rectifier probe changes the AC signal at the test point into an equivalent negative peak DC voltage for application to the emitter follower. The AC voltage monitor can be calibrated in essentially the same manner as explained above for DC voltage calibration. Observe that the AC-voltage tone increments will not have a one-to-one relationship with the DC-voltage tone increments because the rectifier diode in the probe is a nonlinear device.

BASIC AMPLIFIER CLASSIFICATIONS

You will often hear amplifiers described as small-signal or large-signal types, and as low-power or high-power types. From the viewpoint of electrical measurements, these classifications are defined as shown in Figure 2-11. A preamplifier almost always has a small-signal input stage and a large-signal output stage. A power amplifier almost always has a low-power input stage and a high-power output stage.

Figure 2-11

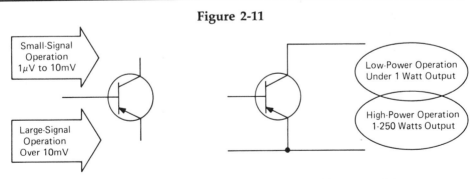

Note: High-fidelity amplifiers differ from public-address amplifiers in that hi-fi units are generally rated for less than 1 percent distortion over a frequency range from 20 Hz to 20 kHz. On the other hand, PA

Figure 2-11 (continued)

units have a considerably higher percentage of distortion and have a smaller frequency range.

Hi-fi amplifiers are also rated for uniformity of frequency response, such as ±1 dB over the rated frequency range. On the other hand, PA amplifiers have considerably less uniformity of frequency response. PA amplifiers are often categorized as utility amplifiers.

Technical Note: "Gung-ho" buffs sometimes make preliminary troubleshooting checks by injecting hum voltage at various points in the circuitry with a screwdriver. This is very poor practice and is a dangerous procedure in the case of high-level power amplifiers. In other words, injection of hum voltage (particularly in the preamplifier circuitry) can cause catastrophic damage to the output section and/or speakers.

Figure 2-11

Basic amplifier classifications. Power ratings are accompanied by distortion ratings.

TROUBLESHOOTING LOW-POWER AUDIO INTEGRATED CIRCUITS

A general purpose low-power audio integrated circuit is generally comprised of a transistor input circuit followed by an IC amplifier. Accordingly, when troubleshooting without service data, we look for a transistor/IC combination. However, in the case of an audio amplifier wherein the noise level is a minor consideration, an IC amplifier will be used without a transistor input section. For example, a stereo-phono amplifier operates with a comparatively high-level signal input, and the inherent noise level of the IC amplifier can be disregarded.

When a stereo-phono amplifier malfunctions, the fault is usually confined to either the L IC chip or to the R IC chip. This duplication of amplifiers makes it easy to close in on the fault through comparison testing (see Figure 2-12). Defective electrolytic capacitors are ready suspects in this type of circuitry. Worn and erratic volume controls are the next most likely culprits. However, if the associated external components "pass inspection," the logical procedure is to replace the IC chip.

A sneaky fault that is occasionally encountered in both mono and stereo IC amplifier systems is inadequate decoupling of the various sections from one another. The result is spurious "audio Q multiplication" with more or less serious distortion. In economy-type systems, the power filter capacitors are the chief means of decoupling. Whenever distortion is accompanied by a noticeable hum level, suspect the power-supply filter capacitors at the outset.

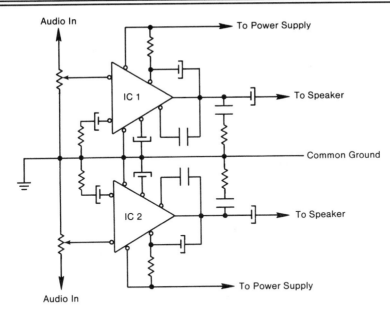

Audio In

To Power Supply

IC 1

To Speaker

Common Ground

IC 2

To Speaker

Audio In

To Power Supply

Note: This type of amplifier is commonly rated for a power output of 1 watt and is provided with switching facilities from FM, AM, and phono units. The phono leads to the switch generally include 1-megohm series resistors for standard frequency compensation.

Figure 2-12

Typical integrated-circuit stereo-phono amplifier arrangement.

IC FUNCTIONAL OVERVIEW

Class AB IC amplifier operation normally occurs with the quiescent (Q) point located below the center of the linear portion on the IC graph. In turn, one-half of the output is a linear reproduction of one-half of the input signal, but the second half of the output signal will be partially absent with respect to the input signal. Distinctions may be made in IC operation between class AB1 and class AB2 modes. Observe that in the class AB2 mode of operation, the Q point is rather close to the cutoff point. However, in the class AB1 mode of operation, the Q point is located about 20 percent or 30 percent above the cutoff point.

Both class AB1 and class AB2 operating modes are encountered in push-pull configurations; generally, cross-over distortion is minimized by operating the push-pull sections back to back. Substantial negative feedback is often employed to further reduce cross-over distortion in IC arrangements that operate in class AB1 or class AB2. Note that the class AB2 mode is somewhat more efficient than the class AB1 mode, but neither has as high operating efficiency as an IC operated in class B. Class AB1 and AB2 modes are used extensively in audio output stages, and also in the specialized ICs used to drive servo motors in, for example, VCR equipment.

When an IC is operated in class B, its Q point is normally located exactly at the cutoff point on the IC graph, with the result that only one-half cycle of the input signal is reproduced in the output from the IC. Except in a few specialized applications, class-B amplifier stages are arranged back to back in push-pull configuration. Cross-over distortion occurs in the basic class-B push-pull arrangement because the input transistors in an IC exhibit barrier potential and do not start conduction until the threshold potential is exceeded (such as 0.5 or 0.6 V). In other words, a "gap" occurs in the collector current through the cutoff region. This is cross-over distortion. Beginners should note that cross-over distortion cannot be corrected by means of negative feedback in this situation because the collector current is zero through the cutoff region, and in turn there is no output signal voltage and no feedback voltage.

However, when the IC is operated in class AB1 or class AB2, more or less collector current flows at all times, and feedback voltage is always present. Accordingly, class AB1 or AB2 cross-over distortion can be minimized by employment of sufficient negative feedback, whereas class-B cross-over distortion is unaffected by any amount of negative feedback.

Next, current amplifiers (emitter followers or linear followers) are basically class-A amplifiers with a voltage gain that may approach but cannot exceed unity. ICs are used in the linear follower mode to function as impedance transformers (high input impedance to low output impedance). From a technical viewpoint, the follower is a current amplifier, and outputs almost the same voltage as the input voltage. Note that in a strict sense, an IC follower should not be termed an emitter follower—only a discrete transistor can operate as a classical emitter follower. Basic parameters are:

1. *Output current.* The rated output current of an IC denotes its application areas and useful range.

2. *Slew rate.* This rating is important in servo-system applications, as in VCRs. IC slew rates are rated up to hundreds of volts per microsecond.

3. *Bandwidth.* The rated bandwidth of an IC in the linear follower mode is the same as in its class-A operating mode.

4. *Full power bandwidth.* The rated full power bandwidth of an IC in the linear follower mode denotes the frequency band over which maximum rated output current is available.

5. *Input resistance.* The input resistance of a linear follower IC is typically in the megohm range.

6. *Output resistance.* The output resistance of a linear follower IC is typically less than 50 ohms. Beginners should note that the rated output resistance is that of the IC itself, and that the effective output resistance is equal to the parallel combination of this rated output resistance and the load resistance that is used.

Next, differential amplifiers (typified by op amps in this case) are ICs that have two input terminals, called the inverting input and the noninverting input. Both input terminals are isolated from the ground by the same impedance (resistance) value. A differential amplifier is usually a class-A voltage amplifier. The IC amplifies the difference in voltage between its two input terminals (it rejects the common-mode voltage applied to its input terminals). The IC is rated for its common-mode rejection ratio. Basic parameters are:

1) *Common-mode rejection ratio (CMRR).* The rated CMRR of the IC is a measure of the amplifier's input characteristic. CMRR ratings of 100 dB or more are available.

2) *Input common-mode range.* This rating of the IC denotes the maximum positive and negative voltage that will be rejected by the CMRR at the input.

3) *Differential voltage gain.* This rating indicates the amplification of the differential input voltage by the IC.

4) *Bandwidth.* This is the rated frequency range for the device when operated conventionally.

5) *Input resistance.* The rated impedance (resistance) between the two differential input terminals.

6) *Input offset voltage.* Usually rated in millivolts; indicates the extent of unbalance between the input terminals of the IC.

The bandwidth of an IC amplifier is a basic concern of the troubleshooter. When substantial feedback is used, the bandwidth is a primary function of the dB feedback. For example, with reference to Figure 2-13, the intrinsic or maximum available gain of the IC in open-loop operation is 100 dB. However, when substantial negative feedback is used (85 dB at zero frequency in this example), the gain of the IC is reduced to 15 dB. Observe that negative feedback action provides an extended high-frequency response, and also provides much more uniformity in frequency response.

Observe also that uniformity of frequency response is dependent on adequate excess loop gain. That is, the excess loop gain is 85 dB at zero frequency, but this excess loop gain eventually becomes zero at the upper frequency limit. We see that as the excess loop gain decreases, the uniformity of amplifier frequency response is lost, and the amplifier frequency response decays to zero

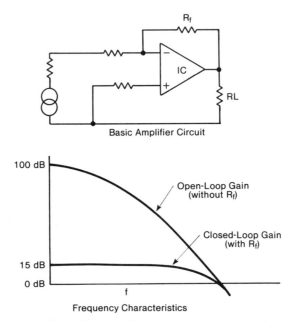

Basic Amplifier Circuit

Note: This basic IC audio amplifier (op-amp) circuit shows how the feedback resistor R_f is connected. In this example, the open-loop gain is 100 dB. Resistor R_f has a value such that the closed-loop gain is only 15 dB. However, the feedback action functions to greatly extend the uniform frequency response of the IC, and to make its frequency response "flat."

Figure 2-13

Basic circuit and typical frequency characteristics for an IC (op amp) as used in audio systems.

at the upper frequency limit. From an applications viewpoint, the IC amplifier is usable out to the frequency where the closed-loop gain begins to drop off noticeably.

TRANSISTOR TESTING WITH THE OSCILLOSCOPE

Some troubleshooters prefer to test transistors with an oscilloscope. Various methods can be used, and the method of preference is dependent on the particular circumstances of the situation. The most basic method is to use a step-function transistor curve tracer with the oscilloscope. In turn, the collector-

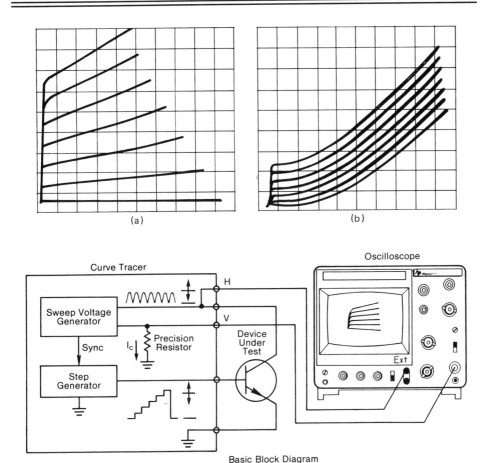

Adapted from releases by B&K Precision, Div. of Dynascan Corp.

Figure 2-14

**Typical collector-family displays: (a) transistor in
normal condition; (b) excessive collector-junction
leakage in transistor.**

family characteristics are displayed, as exemplified in Figure 2-14. Collector-
junction leakage is probably the most common fault, and this malfunction
appears in the photos in Figure 2-14.

Note: The ECG manual available from Sylvania and other manufacturers
provides extensive information on integrated circuit and transistor pinouts, with

typical circuits and typical voltage readings. Quite a few IC problems can be tracked down using this free source of information.

When troubleshooting electronic equipment without service data, you "can't know too much about circuit action." Also, "you can't know too much about reading oscilloscope waveforms." Knowledge of circuit action and observation of trouble clues in distorted waveforms can often save much time that would otherwise be wasted in a "shotgun approach." Knowledge of circuit action involves recognition of equivalent circuits. As an illustration, a circuit comprising a resistor connected in series with a capacitor has an equivalent circuit consisting of an inductor connected in series with a resistor. If the two circuits have the same time-constant, their square-wave or sine-wave responses will be the same.

Because an RC circuit has an equivalent RL circuit, various two-section and three-section RC networks necessarily contain "hidden inductance," and if the Q value of this equivalent inductance is greater than unity, the RC network will "ring" and exhibit overshoot (or undershoot) in its square-wave response. An experienced oscilloscope operator keeps this circuit-action principle in mind when "reading" distorted square-wave response waveforms. Sometimes the distorted waveform will "ring too much," or conversely, it may "ring too little or not at all."

3

PROGRESSIVE AUDIO TROUBLE-SHOOTING PROCEDURES

Stereo Comparison Testing • Testing with Ohmmeters • Quick-Check Comparison Ohmmeter • Automatic Internal Resistance Ohmmeter • Integrated Circuit Preamplifier Considerations • Pointers in Troubleshooting Discrete Preamps • Audio Impedance Checker • Most Significant DC Voltage Measurements • Distortion Case History • Variable Phase-Shift Sourcer; Phase-Angle Quick Checking • Capacitor Leakage Quick Checker • Capacitance Loss and/or Excessive Power Factor Quick Check • Preamp for DC Voltmeter or AF AC Voltmeter • Quick Check for Phase Shift • IC Functional Overview

STEREO COMPARISON TESTING

When troubleshooting without service data, details of the circuitry under test are generally unknown. In turn, it is helpful to start with testing procedures that do not require any knowledge of the normal operating voltages, nor what the function of the circuit under test may be. Comparison testing falls in this category. For example, with reference to Figure 3-1, an AC or DC DVM may be "bridged" between corresponding test points in L and R stereo amplifiers to quickly determine whether any significant voltage difference exists between the two test points.

Comparison testing is sometimes feasible with mono units. Try to locate a similar normally operating mono unit that can be used in comparative testing procedures. If a similar unit cannot be obtained, the troubleshooter must then turn to quantitative measurements, as explained later in the chapter.

TESTING WITH OHMMETERS

Several types of ohmmeters are available for measuring different categories of resistance values. Thus, a simple conventional ohmmeter measures the ohmic value of a resistor or a combination of resistors, the forward or reverse resistance of a semiconductor junction, the leakage resistance in a capacitor, and so on. The applied value of test voltage is not known, and is not of concern in simple situations. This class of ohmmeters is commonly called the high-power (hi-pwr) variety of tester.

Next, a low-power (lo-pwr) ohmmeter typically applies less than 0.1 volt between the points under test. When this ohmmeter is used to measure the value of a resistor, for example, it reads the same value as a hi-pwr ohmmeter. On the other hand, when the lo-pwr ohmmeter is used to measure the junction resistance of a semiconductor device, it always indicates an open circuit (unless the junction happens to be shorted). Many modern DVMs provide a choice of hi-pwr or lo-pwr ohmmeter operation.

QUICK-CHECK COMPARISON OHMMETER

Although the quick-check comparison ohmmeter is unavailable commercially, the troubleshooter can easily devise his or her own quick-check comparison ohmmeter and save a large amount of time in preliminary

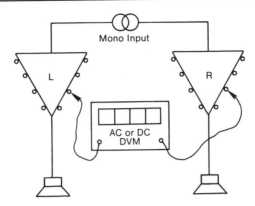

Note: With a mono signal input to the L and R amplifiers, the same AC voltage and the same DC voltage will normally appear at corresponding test points in the amplifiers. In turn, the DVM will normally indicate zero (or nearly zero). If a significant AC or DC voltage is indicated, there is a fault associated with the test point in the malfunctioning unit.

Illustrative example of comparative L and R preamp difference voltages:
  First Preamp Stage
    Collector-to-collector, 0.015 V
    Base-to-base, 0.001 V
    Emitter-to-emitter, 0.002 V
  Second Preamp Stage
    Collector-to-collector, 0.020 V
    Base-to-base, 0.004 V
    Emitter-to-emitter, 0.013 V

Technical Note: If the preamp, driver, and output sections are separate units, be sure that the power is turned off whenever a unit is being disconnected or connected. Otherwise, the output unit may be damaged by a destructive surge voltage.

Figure 3-1

Quick-check method for fault localization in a stereo system.

troubleshooting of circuitry without service data. Figure 3-2 shows a resistance-bridge arrangement used to indicate the inequality of resistance values in the units under test. Although normal component tolerances cause small off-null readouts, a circuit fault will usually result in a large off-zero readout.

In most situations, it is preferable to choose a value for E of approximately 0.1 volt. This provides lo-pwr ohmmeter operation, and normal semiconductor junctions are not turned on. Preliminary tests are therefore made to better advantage because the normal tolerances on semiconductor forward-resistance values are quite large. In turn, the troubleshooter can better judge whether a fault condition is present.

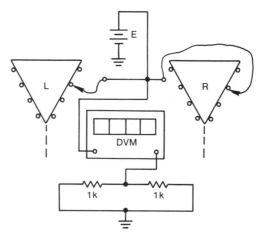

Note: This tester provides quick comparison tests of resistive values in passive circuitry (units turned off). It operates on the unbalanced-bridge principle. If E is a 1.5-volt source, the tester operates as a high-power ohmmeter, and semiconductor junctions in the unit under test will be turned on. Or, if E is a 0.1-volt source, the tester operates as a low-power ohmmeter, and normal semiconductor junctions are "seen" as open circuits. Observe that the chassis of the L and R units are included in the common-ground system. Although this tester can be operated as a quantitative instrument by preparing a chart relating resistance values to unbalanced-bridge voltage values, the primary utility of the tester is that of a qualitative go/no-go indicator.

Figure 3-2

Quick-check comparison ohmmeter arrangement for preliminary troubleshooting without service data.

AUTOMATIC INTERNAL RESISTANCE OHMMETER

Another useful type of ohmmeter that is not available commercially is depicted in Figure 3-3. It operates in active (power-on) circuitry and is a quantitative instrument that measures values of dynamic internal resistance. Semiconductor junctions are, of course, turned on, and are included in the measured value of dynamic internal resistance. However, in normal operation, a semiconductor junction has a close-tolerance resistance value due to its associated bias circuitry.

Much time can often be saved, particularly in "tough-dog" trouble situations, by measuring the dynamic internal resistance of a suspected circuit from various test points. This method is particularly helpful when it is compared with a similar unit that is in normal operating condition. The dynamic internal resistance of a circuit is its "hot" resistance to ground from any chosen test point. This measurement is highly informative because it takes into account not only the fixed resistance values, but also the junction resistances in the circuit under test. Dynamic internal resistance cannot be measured with an ordinary ohmmeter. However, it can easily be measured with the automatic internal resistance ohmmeter depicted in Figure 3-3. It operates as follows:

1. When the test tip is applied at any point in a "live" circuit, the voltage at that point charges the 22-μF capacitor via the 100-kilohm resistor.
2. Then, when the switch is thrown, 1 mA of constant current flows into the test point from the constant-current source.
3. The DVM now indicates the difference between the original voltage (stored in the capacitor) and the change in voltage at the test point resulting from injection of the 1-mA test current.
4. When the DVM is operated on its millivolt range, its readout is equal to the number of ohms of dynamic internal resistance between the test point and ground via the circuit under test.

A practical 1-mA constant-current source for use in checking typical semiconductor circuitry can be devised by connecting a 100,000-ohm resistor in series with the output from a 100-volt power supply, as shown in Figure 3-3. (Your workbench DC power supply is handy for this purpose, provided that it is capable of outputting 100 volts.) This is a practical constant-current source because most bipolar transistor circuitry has an internal resistance considerably less than 100,000 ohms.

Observe that the arrangement shown in Figure 3-3 is polarized for checking at a negative test point. If you wish to check dynamic internal resistance at a positive test point, the polarity of the constant-current source should be reversed. Note that the 22-μF capacitor will proceed to charge slowly from the constant-current source when the switch is thrown. This is just another

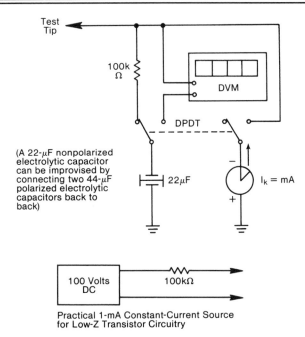

Practical 1-mA Constant-Current Source
for Low-Z Transistor Circuitry

Note: The capacitor functions as a charge-storage device to hold the original voltage value which was present at the node under test. Then, when the switch is thrown, 1 mA of constant current is forced through the dynamic internal resistance. This produces a voltage increment of I_kR_{in}, or $0.001R_{in}$ volt. Thus, the DVM indicates the difference between the stored voltage value and its incremented value. Since $I_k = 1$ mA, the DVM millivolt readout equals the value of the dynamic internal resistance in ohms. (I_k may consist of a 100-kilohm resistor and a 100-volt power supply.) The DPDT switch may be a miniature toggle type such as the Radio Shack 275-664.

Figure 3-3

Arrangement of a dynamic internal resistance ohmmeter.

way of saying that the initial readout on the DVM will gradually decay toward zero. In turn, the technician should observe the DVM readout as soon as the switch is thrown.

INTEGRATED CIRCUIT PREAMPLIFIER CONSIDERATIONS

When you troubleshoot IC preamplifiers without service data, it is helpful to keep the basic network features in mind. A simple arrangement was shown previously in Figure 1-2. A more elaborate arrangement is depicted in Figure 3-4. This network includes bass and treble tone controls and an L/R balance control in addition to the volume control. Signal progression takes place from the input connector to the tone controls via the amplifier devices inside the chip. From this point, the signal proceeds to the volume and balance controls, and again to amplifier devices in the chip. Finally, the signal is fed to the output connector.

Accordingly, the signal level is normally greater at the input to the tone controls than at the input connector. Similarly, the signal level is normally greater at the output connector than at the tone controls. In case the trouble symptom is "weak output," it is necessary to determine whether the fault is in the integrated circuit, or in its associated external circuitry. In the absence of specific clues, electrolytic capacitors should be checked first. A worn and/or erratic volume control is also a ready suspect. If the external circuitry "passes inspection," it can be logically concluded that the IC is faulty.

As shown in Figure 3-5, the maximum available voltage gain of a typical preamplifier is over 2500 times, or 69 dB (without taking the input and output impedances into account). Both IC and discrete preamps commonly include three stages (input, driver, and output stages). A substantial amount of negative feedback is provided in a hi-fi preamp to maintain distortion at less than 1 percent. When the trouble symptom is "excessive distortion," capacitors and resistors in feedback loops are ready suspects.

POINTERS IN TROUBLESHOOTING DISCRETE PREAMPS

An integrated-circuit preamp is often preceded by a discrete input stage, or it may consist of discrete circuitry throughout. Nearly all discrete stages normally operate with all transistor terminal voltages positive with respect to ground, or with all transistor terminal voltages negative with respect to ground. Most stages are operated in the common-emitter mode. If a stage is operated in the emitter-follower mode, its transistor terminal voltages will also be either all positive or all negative with respect to ground.

An exception to the foregoing rule occurs in the case where a stage is operated in the common-base mode. In this situation, the emitter will normally have opposite polarity to that of the collector. However, the common-base

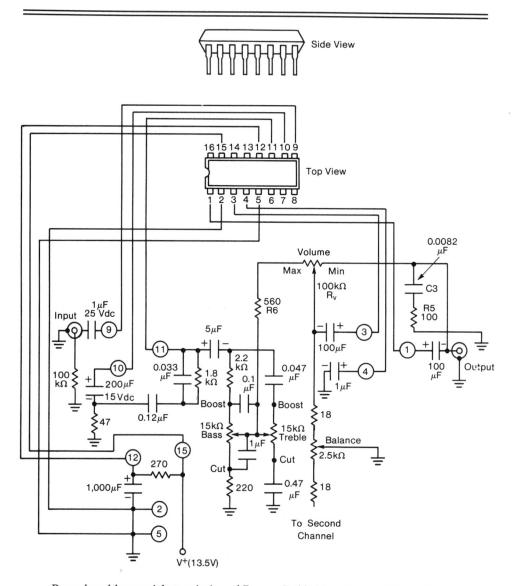

Reproduced by special permission of Reston Publishing Co. and Derek Cameron from Handbook of Audio Circuit Design.

Figure 3-4

Typical integrated-circuit preamplifier arrangement.

Figure 3-5

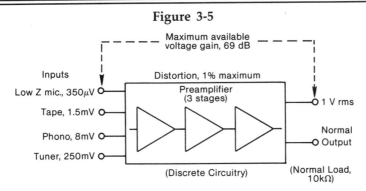

Note: The first stage in a preamplifier commonly operates with comparatively low collector voltage to minimize circuit noise (low collector voltage is not necessarily an indication of trouble). The third stage in a preamp is often operated as an emitter follower; it provides no voltage gain, but has power gain and provides impedance transformation.

Technical Note: In the case of direct-coupled circuitry, this arrangement is a basic example of deceptive circuitry, insofar as DC voltage measurements are concerned. In other words, a very small off-value of bias voltage in the first stage appears as a large off-value in the second stage, and as a very large off-value in the output stage. This is just another way of saying that if no device or component defect is found in a suspected stage, proceed to check the preceding stage.

Reproduced by special permission of Reston Publishing Co. and Derek Cameron from Handbook of Audio Circuit Design.

Figure 3-5

**Hi-fi audio preamps typically include three stages,
with maximum voltage gain approaching 70 dB.**

configuration is rarely encountered in preamp circuitry. By way of comparison, output-amplifier circuitry employs common-base stages in various subsections.

The essential point to keep in mind when you troubleshoot bias circuitry in a preamp is:

1. As shown in Figure 3-6, transistors have barrier potentials, and the collector current is normally zero if the base-emitter voltage is less than the barrier potential. The barrier potential for a normal germanium transistor is approximately 0.2 volt, and is about 0.55 volt for a normal silicon transistor.

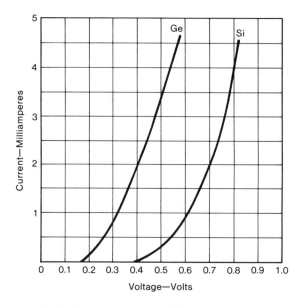

Note: Because of the barrier potential that exists in the base-emitter circuit, a transistor is normally reverse-biased by a fraction of a volt if the base terminal is shorted to the emitter terminal. This inherent reverse bias normally stops collector-current flow and is the basis of the turn-off quick check.

Technical Note: The precise barrier potential is somewhat different for various designs of transistors; the barrier potential is a function of temperature, and is specified for room temperature (25°C).

Figure 3-6

Junction current does not flow in a semiconductor device unless the forward bias voltage exceeds the barrier potential.

AUDIO IMPEDANCE CHECKER

A highly useful audio quick-checker is shown in Figure 3-7. This is an audio impedance checker that is used in preliminary troubleshooting procedures. It does not measure AC impedance in ohms—instead, it compares the AC impedance at a test point in a bad amplifier with the AC impedance at a corresponding test point in a similar good amplifier.

A practical example of application for the audio impedance checker is also shown in Figure 3-7. This quick checker is particularly useful for spotting open capacitors, inasmuch as they escape DC voltage and resistance measurements. Note that a capacitor does not have to be completely open in order to be "caught" by the audio impedance checker. For example, if an electrolytic capacitor has lost half of its capacitance, or if it has developed an abnormally high

Figure 3-7

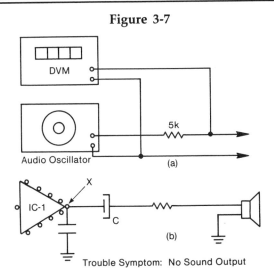

Trouble Symptom: No Sound Output

Procedure: The DVM is operated on its AC voltage function. The audio oscillator is set to 1 kHz with an output level of 2 or 3 volts, as indicated when the output leads are open-circuited.

Practical Note: The peak AC voltage from the test point to ground should not exceed 500 mV, in order to avoid junction turn-on and possible confusion of test results.

This is an example of an AF integrated circuit driving a speaker through an electrolytic capacitor and a resistor. (An audio impedance checker is also valuable for testing other types of electronic circuitry.)

Figure 3-7 (continued)

In this example, there was no sound output from the speaker. All of the DC voltages at the IC terminals made sense on a comparative basis. However, when the audio impedance checker was applied at point X, the DVM indicated that the impedance was much higher in the "bad" amplifier than in the "good" amplifier:

"Bad" Amplifier—56 mV
"Good" Amplifier—5 mV
(DVM indicates 2.75 V on open circuit)

The comparatively high AC voltage reading in the "bad" amplifier indicated that the impedance to ground from point X was abnormally high. In this type of circuitry, electrolytic capacitors are ready suspects.

In turn, capacitor C fell under suspicion. When a signal voltage was next applied to the input of IC-1, an AC voltmeter showed a signal present on the left-hand end of C, but zero signal on the right-hand end of C.

Therefore, the troubleshooter concluded that C was open-circuited. When a test capacitor was bridged across C, the speaker resumed sound output.

Figure 3-7

Audio impedance checker: (a) test setup; (b) example of application.

power factor, the associated circuit impedance will be changed, and an audio impedance checker will show this change in a comparative quick test.

MOST SIGNIFICANT DC VOLTAGE MEASUREMENTS

Although all of the DC voltages in a circuit have some relationship with circuit action, the collector-to-ground voltage is most significant in preliminary troubleshooting procedures. A practical example is shown in Figure 3-8. This is a basic CE configuration with voltage and current feedback, and with collector bias. Two common trouble conditions are considered in this example: (1) serious collector-junction leakage, and (2) open collector junction. DC voltage measurements may be made from base to ground, from emitter to ground, and from collector to ground.

As listed in Figure 3-8, the base-to-ground voltage changes slightly in this example from its normal value to the trouble values: 0.55 to 0.556 to 0.522 V.

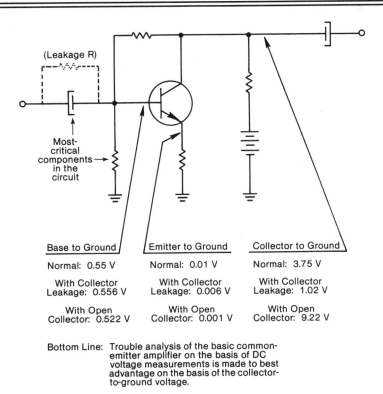

(Leakage R)

Most-
critical
components →
in the
circuit

Base to Ground	Emitter to Ground	Collector to Ground
Normal: 0.55 V	Normal: 0.01 V	Normal: 3.75 V
With Collector Leakage: 0.556 V	With Collector Leakage: 0.006 V	With Collector Leakage: 1.02 V
With Open Collector: 0.522 V	With Open Collector: 0.001 V	With Open Collector: 9.22 V

Bottom Line: Trouble analysis of the basic common-emitter amplifier on the basis of DC voltage measurements is made to best advantage on the basis of the collector-to-ground voltage.

Figure 3-8

Practical example of DC voltage changes caused by common defects in an amplifier stage.

Similarly, the emitter-to-ground voltage changes slightly from its normal value to the trouble values: 0.01 to 0.006 to 0.001 V. On the other hand, the collector-to-ground voltage changes greatly from its normal value to the trouble values: 3.75 to 1.02 to 9.22 V.

Every troubleshooter occasionally encounters the necessity (or desirability) of opening a circuit for testing. A familiar example is a bypassed emitter resistor when the emitter voltage is subnormal and the resistor measures only a fraction of its normal resistance value. It is impossible to decide at this point whether the emitter resistor is faulty or whether the bypass capacitor is faulty (leaky). To make this distinction, it is necessary to open the circuit between the capacitor and the resistor. Ordinarily, the most practical procedure is to make a razor slit in the PC conductor that connects the capacitor to the resistor. Then, conclusive tests can be made to determine which component is faulty. After the defective component is replaced, the PC conductor is then repaired with a small drop of solder. (Slitting the foil is a "last resort" procedure.)

Another example is suspected leakage in a coupling capacitor. Referring to Figure 3-8, leakage in the input coupling capacitor can cause an abnormally high bias voltage or a subnormally low bias voltage, depending on the voltage level of the preceding circuitry. In the case of the output coupling capacitor, leakage will usually result in a subnormally low collector voltage. It is generally difficult to conclude whether off-value voltages are being caused by coupling-capacitor leakage, or by other faults such as off-value resistors or transistor defects. Here again, the most practical procedure in most cases is to make a razor slit in the PC conductor that connects the capacitor to its neighboring circuit.

DISTORTION CASE HISTORY

Observe the complementary-symmetry stage exemplified in Figure 3-9. The trouble symptom under analysis was distortion—this distortion was most

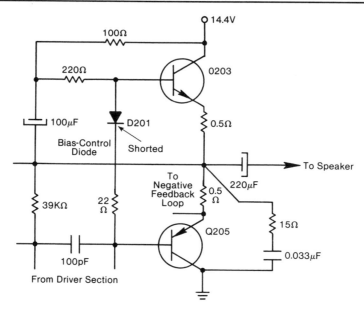

Note: This is a low-power amplifier because it is rated for 0.9 watt power output. (High-power amplifiers are rated for more than 1 watt power output.)

Figure 3-9

Case history of distortion caused by a shorted diode.

objectionable at low volume levels, and was less severe at higher volume levels. DC voltage measurements at the transistor terminals turned up the following facts:

1. The base-emitter voltage on Q205 would normally be in the order of 0.6 V, but measured only a small fraction of a volt. (It was evident that the distortion was bias-related.)

2. When the bias voltage was monitored with a program signal inputted, it was observed that the bias voltage varied with the signal level. At times, the bias-voltage polarity would reverse.

3. A check of the voltage drop across bias diode D201 showed zero volts— it was apparent that the diode was shorted. Replacement of the diode restored the stage to normal operation.

CAPACITOR LEAKAGE QUICK CHECK

When troubleshooting without service data, quick checks that require no attention to circuit details are preferred to analyses that require calculation of circuit action. For example, it is desirable to have a quick-check method available to determine whether a coupling capacitor, a bypass capacitor, or a decoupling capacitor may be leaky. The following quick check for suspected capacitor leakage does not require knowledge of resistance or capacitance values, nor of the circuitry associated with the capacitor under test. The quick check is based on the following principles:

1. If a capacitor is not leaky, and a DC voltage is applied across the capacitor, there is zero DC current in the conductor that connects the capacitor to its associated circuitry.

2. If a capacitor is leaky, and a DC voltage is applied across the capacitor, there is more or less DC current in the conductor that connects the capacitor to its associated circuitry.

3. When there is DC current in a conductor, there is at least a slight voltage drop from end to end of the conductor due to its inherent resistance.

4. A sufficiently sensitive DC voltage indicator will show whether DC current is or is not present in the conductor.

Figure 3-10 shows the lead to the coupling capacitor being checked for an IR drop by touching the test leads of the mini-amp to separated points along the coupling-capacitor lead. A mini-amp such as the Archer (Radio Shack) 277-1008 is suitable for this purpose. As a practical consideration, it is advisable to check the longer of the two capacitor leads so that the test points will be separated as much as possible. A greater separation of the test points provides greater sensitivity of indication. This quick check is useful for any of the capacitors in an audio system, because a capacitor in normal condition conducts zero direct current.

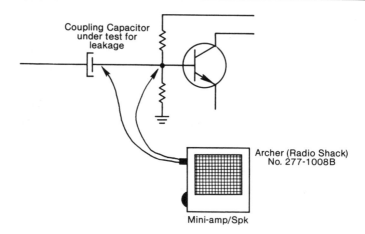

Note: This is a useful quick check for capacitor leakage. The amplifier leads are connected at separated points on the lead from the capacitor to its following circuitry. The coupling capacitor is normally an open circuit to DC, so that there is no current in the lead although DC voltage is present. If there is no leakage in the capacitor, there will be no click from the speaker when the test lead is touched to the capacitor lead. On the other hand, if there is leakage in the capacitor, a click will be heard from the speaker. The mini-amp volume control is turned up to maximum for testing. Although the right-hand lead to the coupling capacitor is being tested in the diagram, it is equally informative to test the left-hand lead to the coupling capacitor.

Technical Note: Observe that the coupling capacitor in the mini-amp/spk must be in good condition. Otherwise, leakage current from this capacitor will confuse the test results.

Figure 3-10

Quick check for electrolytic capacitor leakage with
a high-gain mini-amp/spk.

CAPACITANCE LOSS AND/OR EXCESSIVE POWER FACTOR QUICK CHECK

Electrolytic capacitors used for coupling, decoupling, or bypassing action often develop loss of capacitance and/or excessive power factor with age. In turn, the defective capacitor does not represent a virtual short-circuit for AC, but instead presents substantial reactance or impedance to signal flow, with comparatively large attenuation of low frequencies. When this fault is suspected, a quick check can easily be made as shown in Figure 3-11. An AC DVM is connected across the capacitor, and a 20-Hz test signal is applied to the amplifier. In turn, if a significant signal voltage is indicated on the DVM, the troubleshooter concludes that the capacitor is defective. (The rated frequency response of a hi-fi amplifier is from 20 Hz to 20 kHz.)

PREAMP FOR DC VOLTMETER OR AF AC VOLTMETER

Although modern service-type voltmeters are comparatively sensitive, the troubleshooter occasionally needs somewhat higher sensitivity in low-level tests. An instrument preamp can be readily constructed as explained below, and will provide an indication sensitivity increase of 100 times. For example, when used with a 20,000 ohms-per-volt VOM, this preamp provides a sensitivity of 50 microvolts. When used with a 50,000 ohms-per-volt VOM, it provides a sensitivity of 25 microvolts.

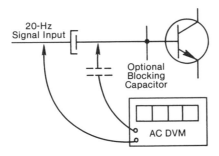

Note: This quick check shows whether a significant signal voltage drop is present across the coupling capacitor at its lowest rated frequency of operation. If the capacitor is in normal condition, practically no signal voltage will be dropped. If the AC DVM does not have a built-in blocking capacitor, an external blocking capacitor should be used to avoid false indication due to the DC voltage drop across the coupling capacitor.

Figure 3-11

Significant signal voltage is dropped across the coupling capacitor if it has deficient capacitance or a high power factor.

A general-purpose voltmeter preamp can be constructed from a Radio Shack Type 741 operational amplifier. This integrated circuit has the package pinout shown in Figure 3-12. It provides a stable DC and audio-frequency amplifier with a voltage gain of 100 times when connected as shown in Figure 3-13. The IC is powered by two 9 V batteries. Note the following points:

1. The voltage gain of the amplifier is equal to the ratio of the feedback resistor value to the input resistor value (1,000,000/10,000). It is recommended that ±1 percent resistors be used to obtain precisely 100× amplification.

2. Pin 2 of the IC is a "virtual ground" due to the substantial amount of negative feedback that is provided. In turn, the input resistance of the amplifier is 10,000 ohms.

3. The output impedance (resistance) of the op amp is approximately 75 ohms. Accordingly, the op amp can be used to drive any VOM, regardless of its ohms-per-volt rating.

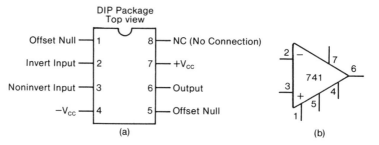

(a)   (b)

Note: The Type 741 op amp has a maximum available gain of 200,000 times. However, it is always operated with substantial negative feedback, so that the MAG is reduced to 100 times, for example. The op amp is powered by two batteries; the $-V_{cc}$ battery has its positive terminal connected to ground, and the $+V_{cc}$ battery has its negative terminal connected to ground. In most applications the input voltage is applied between the inverting input (pin 2) and ground. Output is taken from pin 6 to ground.

Figure 3-12

Operational amplifier, Type 741: (a) package pinout; (b) symbol.

Figure 3-13

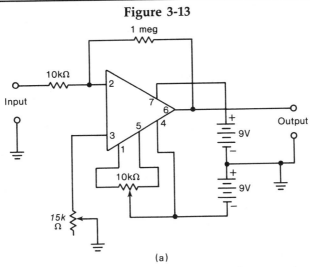

(a)

Note: Unless nulling adjustments are provided, the output terminal of the op amp will not rest exactly at zero. This is called the offset voltage. The offset voltage is precisely cancelled out by adjustment of the 15k and 10k potentiometers.

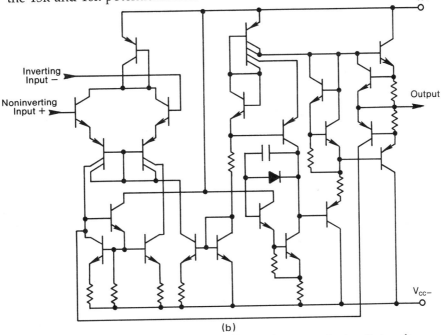

(b)

Note: The question of what is inside an integrated circuit is often asked. This typical schematic diagram shows that the IC package

Figure 3-13 (continued)

contains transistors, diodes, and resistors. Differential amplifiers are
employed. The input stage has an inverting input and a noninverting
input; it is called the differential stage. The intermediate stage is
called the level-shifting amplifier. The output stage is called the output amplifier; it has one output terminal. Observe that the op amp
has an extremely high input impedance, and a very low output impedance. However, it is almost always used with a large amount of
negative feedback, with the result that its effective input impedance
becomes very low.

Technical Note: This example of IC circuitry illustrates an important
point from the viewpoint of the practical troubleshooter—that is, it is
far more difficult to follow integrated circuit action than to follow
discrete circuit action. Accordingly, when troubleshooting in integrated arrangements, the technician is primarily concerned with comparative voltage and resistance values, without regard to the detailed
circuit actions that are involved.

Figure 3-13

**Type 741 op-amp arrangement: (a) configuration
for the 100 × voltmeter preamp; (b) internal
circuitry of the op amp.**

To zero the preamp output, short-circuit the input terminals and make
nulling adjustments as follows:

1. Start with the 15 K potentiometer set to 10,000 ohms or slightly more.
2. Adjust the 10 K potentiometer to obtain a zero readout on the voltmeter.
3. If a precise zero readout is not obtained, touch up the adjustment of the
   15 K potentiometer.

These nulling adjustments are quite critical, and you may find it desirable to provide vernier adjustments. For example, a 1 k potentiometer may be
connected in series with the 15 K potentiometer. Similarly, a 1 k potentiometer
may be connected in series with the 10 k potentiometer. If you have difficulty in
nulling out the last vestiges of offset voltage, a final zeroing can be obtained by
inserting a millivolt bias box in series with the ground lead of the voltmeter. A
basic millivolt bias box is configured as shown in Figure 3-14.

Consider voltage measurements when the op amp is connected to a
service-type DVM that indicates millivolt values. The op amp increases the
indication sensitivity by 100 times, so the DVM indicates a voltage level as low as

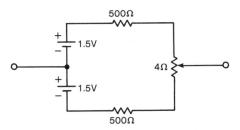

Note: Although this bias arrangement inserts approximately 250 ohms of resistance in series with the DVM's ground lead, the accuracy of voltage measurement is practically unaffected inasmuch as the input resistance of the DVM is 10 megohms.

Figure 3-14

Basic millivolt "bias box" arrangement for final nulling of the preamp output offset.

10 microvolts. In other words, a 1-mV readout corresponds to a 10-μV input voltage to the op amp.

Troubleshooters do not expect a DVM to always automatically null out precisely on each of its ranges. For example, a professional DVM might normally indicate 0.1 mV (instead of 000.0) when its test leads are short-circuited together. A service-type DVM might indicate 1 or even 2 mV (instead of .000) when its test leads are short-circuited together.

In other words, a DVM may normally display a small offset voltage when its test leads are short-circuited together. However, the DVM can be precisely zeroed in any case where it is used with a preamp as described above. The nulling procedure not only cancels out any slight residual preamp offset, but also cancels out any DVM offset.

QUICK CHECK FOR PHASE SHIFT

It is sometimes helpful to know whether phase shift has occurred from one point in an audio network to another point. A handy quick checker for this purpose is shown in Figure 3-15. It consists of a pair of mini-amps such as the Archer (Radio Shack) 277-1008. The phase check is made with L and R earphones, so that the left ear of the operator hears the sound from the L test leads, and the right ear hears the sound from the R test leads. Relative phase checks can be made in the same circuit or in different circuits, in mono audio systems, or in stereo audio systems.

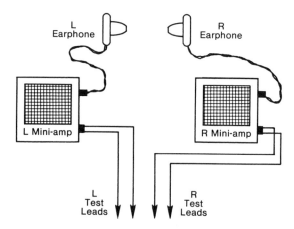

Note: This phase quick checker operates on the basis of directional judgments in the binaural hearing process. If the signal applied to the L test leads has the same phase as the signal applied to the R test leads, the listener judges that the source of the sound is directly in front of him or her. On the other hand, if the signal applied to the L test leads has a different phase from the signal applied to the R test leads, the listener judges that the source of the sound is to his or her left or right by an amount that depends on the difference in phase. The volume controls on the L and R mini-amps should be adjusted for a comfortable listening level, and for about the same loudness from each of the earphones.

Figure 3-15

Quick checker to determine whether two audio signals are in phase, or out of phase with each other.

The binaural hearing process makes directional judgments on the basis of relative phase with respect to any given audio frequency. Thus, if the input signal to the L test lead has a certain phase, and the input signal to the R test lead is progressively shifted in phase through a 360-degree range, the operator will judge that the source of the sound is moving around him or her in a circle. This is essentially a qualitative tester and it does not provide precise judgment of phase difference.

Note that the No. 277-1008 mini-amp has 5 kilohms input resistance and does not include an input blocking capacitor. If objectionable DC drain-off would occur in a test, blocking capacitors should be included in series with the test leads to the mini-amps. If a follow-up test is desirable to measure the phase

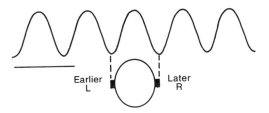

Earlier
L

Later
R

Note: As shown in the diagram above, sound that comes from the left of the listener has a phase that leads at the left ear, and lags at the right ear. In the hearing process, this leading phase is automatically interpreted as a left-hand source. Conversely, sound that comes from the right of the listener has a phase that leads at the right ear, and lags at the left ear. In the hearing process, this leading phase is automatically interpreted as a right-hand source. It follows that the quick checker depicted in Figure 3-15 employs this same principle. In other words, if the sound seems to come from the left of the troubleshooter, the left-ear signal phase is leading the right-ear signal phase. Conversely, if the sound seems to come from the right of the troubleshooter, the right-ear signal phase is leading the left-ear signal phase.

Figure 3-16

Sound coming from the left of the listener has a phase that leads at the left ear and lags at the right ear.

angle between the two signal sources, a phase-shift sourcer and DVM should be used as explained previously.

As shown in Figure 3-16, the troubleshooter can determine whether the right-ear signal phase leads or lags the left-ear signal phase, based on the apparent direction of the sound source. That is, if the sound seems to be coming from the left, then the left-ear signal source is leading the right-ear signal source. Or, the right-ear signal source is lagging the left-ear signal source. Insofar as circuit action is concerned, note that in a series RC circuit the voltage across R leads the source voltage, and the voltage across C lags the source voltage.

IC FUNCTIONAL OVERVIEW

An integrated-circuit operational amplifier (op amp) is basically a DC-coupled, high-gain differential amplifier. It is almost always used with a large amount of external negative feedback. (See Figure 3-17.) The IC, per se, has an

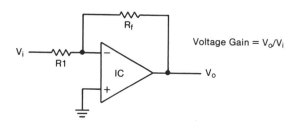

Note: This basic amplifier arrangement, wherein the IC is operated with a large amount of negative feedback, normally has a voltage gain that is almost exactly equal to R_f/R_1. Thus, if R_1 = 1000 ohms, and R_f = 100,000 ohms, the amplifier gain is normally equal to 100. The input resistance of the amplifier is almost exactly equal to R_1 (1000 ohms in this example). Observe that these functional relations imply that the input resistance to the inverting input of the IC is almost zero. This is why the IC input terminal is called a virtual ground in this type of amplifier circuit. Note that the output resistance of the IC is very low, with the result that the foregoing characteristics are essentially independent of the load-resistance value that may be used.

Figure 3-17

Functional overview of an IC amplifier operated with a large amount of negative feedback.

extremely high gain; this gain is called its *open-loop gain* and it is measured in the absence of any negative feedback. In open-loop operation, the op amp has extremely high input impedance, and very low output impedance. Its frequency response is typically limited to the audio range in open-loop operation.

When the IC is operated with significant negative feedback, the op amp provides a fixed gain versus frequency, which is almost exactly determined by the ratio of series input resistance to feedback resistance in the amplifier circuit. *In this mode of operation, the input resistance is almost zero (a virtual ground) due to the cancelling action of the feedback voltage.* Also, the frequency range of the amplifier is greatly extended by negative feedback action. Occasionally, we find an op amp IC operated with zero feedback resistance (inverting input connected directly to the output terminal). Unity gain is developed in this mode of operation.

The IC also can be operated with a high value of feedback resistance, and in turn, the gain of the op amp is increased. Its bandwidth is decreased, as anticipated. An op amp always has maximum open-loop gain at zero frequency, and its frequency response drops off progressively as the operating frequency is increased. At some point, the op amp will have unity gain. When used with negative feedback, the difference between the closed-loop gain and the corresponding open-loop gain at a given frequency is called the *excess loop gain*. Excess loop gain is, of course, greatest at zero frequency and drops off progressively as the operating frequency is increased.

From the applications viewpoint, the foregoing variation in excess loop gain limits the practical operating frequency range of the IC. Stated another way, the "penalty" of low excess loop gain is development of phase shift and development of noticeable distortion in the frequency range associated with low excess loop gain. It will be seen that if excessive phase shift occurs due to low excess loop gain, the intended negative feedback voltage then becomes positive feedback voltage, and the op amp "takes off" and oscillates uncontrollably.

To obtain a greater effective frequency range without oscillation, many op amps provide for connection of an external phase-shifting capacitor to minimize development of phase shift in the region of low excess loop gain. Of course, if this capacitor becomes defective, the IC will then break into violent oscillation. Basic functional parameters are:

1. *Open-loop gain.* The open-loop gain for the IC is usually rated for DC; a graph of open-loop gain versus frequency is often also supplied and can be of importance in many applications. The value of the open-loop gain is an indication of the degree of stability that can be obtained in circuits with negative feedback.

2. *Input impedance.* The rated input impedance is accurately represented in terms of an equivalent RC parallel circuit shunted across the IC input terminals. When low-frequency operation only is of concern, the input impedance may be rated simply as an input resistance.

3. *Input offset voltage.* The input offset voltage is measured by using a feedback resistance that provides a large value of fixed gain. The value of the error measured at the output is then divided by the gain figure. Input offset voltage is the result of residual unbalances in the differential stages inside the IC. In most applications, the input offset voltage can be nulled out by means of a potentiometer, as explained in the text. Note that a typical op amp has three internal stages: the differential stage, the level-shifting stage, and the output stage. Differential circuitry is employed throughout.

4. *Input noise.* Input noise ratings are frequently given as a signal-to-noise ratio, a noise figure, or input noise current; low-frequency noise and wideband noise are rated separately. Spot noise occurs at certain frequencies and is rated in detailed manufacturers' specifications.

5. *Common mode rejection ratio (CMRR).* This rating is the same as described previously for differential amplifiers.

6. *Maximum common mode voltage.* Same as CMRR.

7. *Drift versus temperature.* The temperature change in offset voltage, bias

current, difference current, and related parameters is rated as the temperature coefficient for each parameter, or it may be listed in tabular form or shown as graphs.

8. *Bias current.* The bias current rating is the current from an infinite source impedance (constant-current source), which, when applied to either input terminal, will drive the output to zero.

9. *Difference current.* The rated difference current for the IC is the difference between the inverting input terminal current and the noninverting input terminal current required to drive the output to zero.

10. *Bandwidth.* Denotes the frequency range over which various specifications apply.

11. *Full power response.* The rated sine-wave output at the maximum frequency that provides unity closed-loop gain for a specified percentage distortion in the rated output load.

12. *Rated output voltage.* The rated minimum peak output voltage at the rated current for completely linear operation.

13. *Rated output current.* The rated minimum output current at the rated output voltage.

14. *Overload recovery.* The rated time that is required for the output voltage to return to the rated value after a 50 percent overdrive saturation occurs.

15. *Settling time.* The rated period from the input of a step voltage to the time that the output voltage reaches the specified error range of its final value. This rating is usually based on unity gain and without capacitive loading.

16. *Slew rate.* Usually specified in volts per microsecond; it denotes the maximum rate of change in output voltage for an input step-voltage response.

Step-voltage response can be effectively checked only with an oscilloscope with respect to a square-wave input. Note that troubleshooting without service data is often facilitated by checks of phase distortion in addition to amplitude distortion. When a coupling or decoupling capacitor is marginal, its deficiency becomes apparent as signal phase shift before it deteriorates to the point that its deficiency becomes apparent as amplitude distortion. Leading or lagging phase shift in square-wave response is exhibited as "tilt."

4

TROUBLE-SHOOTING RADIO RECEIVERS WITHOUT SERVICE DATA

Preliminary Considerations • "The Sound of Troubles" • Useful Old and New Quick Checks • Tuned IF Signal Tracer • AM Radio Receiver Layouts • Widely Used AM Radio Receiver Transistor Arrangements • Preliminary Alignment Procedure • Alignment with Test Equipment • Stage Identification in Dead Circuitry (Resonance Probe) • High-Impedance Tuned Signal-Tracing Probe • Autodyne Circuit Troubleshooting • Impedance Check at Battery Clip Terminals • FM/AM Radio Receiver Troubleshooting • Capture Effect • Identification of FM Transformers in FM/AM Receiver • "The Sound of Full Rectification" • IC Functional Overview • Sweep Alignment with the Oscilloscope

PRELIMINARY CONSIDERATIONS

Troubleshooting radio receivers without service data requires one of two basic approaches, depending on whether the radio receiver is workable (although malfunctioning), or whether it is unworkable (dead). Thus, when the receiver is workable, although malfunctioning, preliminary analysis consists in "sizing up" the trouble symptoms for clues concerning the nature of the fault. Details of this procedure are explained further on the following pages.

When the receiver is unworkable (dead), the power-supply voltage should be checked first. If the supply voltage is normal, the troubleshooter proceeds to make a series of quick checks to determine which sections of the receiver are operable, and where the signal is being stopped. Some of these quick checks are simple "commonsense" procedures, and others require various kinds of test equipment. These procedures and requirements are described and illustrated below. (Note that some troubleshooters feel that it doesn't pay to repair an inexpensive radio.)

"THE SOUND OF TROUBLES"

When troubleshooting radio receivers without service data, it is helpful to make full use of techniques that do not require any knowledge of the circuitry under test, nor of the sectional layout of the receiver. For example, when the receiver is workable, although malfunctioning, considerable time and effort can often be saved if the troubleshooter knows how to recognize "the sound of troubles."

With the exception of the smallest receivers, the output section operates in push-pull, either in class B or in class AB. One of the common failures is a "dead" output transistor. This malfunction leaves only one-half of the push-pull stage operative, and it is accompanied by "a sound of trouble" that is recognized immediately by the experienced troubleshooter. In turn, the faulty stage can be "spotted" at the outset.

A familiarization experiment to acquaint the troubleshooter with the "sound of class B" is shown in Figure 4-1. Basically, the arrangement provides half-wave rectification of a radio receiver's sound output signal. The "sound of class B" is harsh compared with the normal timbre of the signal. The beginner should make a mental note of this harsh timbre, so that it will be recognizable when a malfunctioning receiver is "sized up." For comparison, the 1N34A diode may be shorted-out in this experiment. In turn, the normal timbre of the signal is reproduced, compared with the "sound of class B."

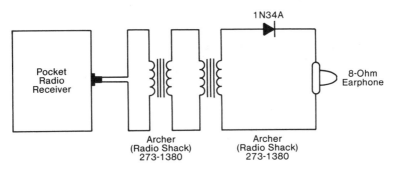

Note: This is a familiarization experiment to demonstrate "the sound of class B." It uses a rectifier diode in series with an earphone. The output from the radio cannot be used directly (in most cases) because there is a small DC component present that would impair the operation of the 1N34A. Accordingly, a pair of audio output transformers is required to isolate the diode, and also to provide suitable impedance matching. The radio drives the 8-ohm primary of the first transformer, and its 1-k secondary is connected to the 1-k secondary of the second transformer. In turn, the 8-ohm winding of the second transformer is connected to the diode and earphone.

Observe that sound output from the earphone has a characteristic "scratchy" and "crackly" or "rattly" distortion, although speech can usually be understood.

Figure 4-1

An arrangement to illustrate the "sound of class B."

A modification of the foregoing experimental arrangement which illustrates the *"sound of class AB"* is shown next in Figure 4-2. Here, the 1N34A diode is shunted across the 1 k windings of the audio transformers. Due to the barrier potential of the diode, less than one-half of the positive excursion of the signal is removed. In turn, a class-AB signal is applied to the earphone. Observe that at low-volume levels in this experiment the timbre of the signal sounds normal. This is because the barrier potential of the diode has not been exceeded.

Clipping is another malfunction that may be encountered in any type of radio receiver. A further modification of the foregoing experimental arrangement to demonstrate the *"sound of clipping"* is depicted in Figure 4-3. (A second diode is shunted across the 1-k windings of the audio transformers in reverse polarity to the first diode.) As before, it will be observed that at low-volume

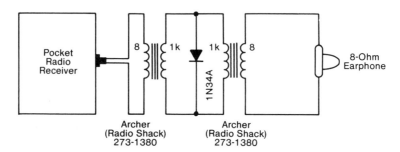

Note: This is another familiarization experiment to demonstrate "the sound of class AB." It uses a diode as described previously for class-B operation, but in this case the diode is shunted across the 1-k windings of the audio transformers. In turn, class-AB operation is obtained at moderate volume levels. Observe that the "sound of class AB" is somewhat less harsh than "the sound of class B." Musical passages are noticeably distorted, and speech is also distorted, although it is usually intelligible.

Figure 4-2

Another experimental arrangement to illustrate
"the sound of class AB."

levels where the barrier potentials are not exceeded, the timbre of the signal sounds normal. As the volume level is increased, characteristic spurious signal components and increased noise output become evident. Observe that clipping distortion is less objectionable than half-wave distortion (except at comparatively high volume levels).

Subnormal bandwidth is sometimes encountered in a malfunctioning radio receiver. It is usually localized to the IF section, inasmuch as most of the gain and selectivity in a receiver are provided by the IF section. An experimental arrangement that demonstrates the *"sound of insufficient bandwidth"* is shown in Figure 4-4. Here, a gimmick consisting of a short clip lead is run from the input end of the detector diode to the vicinity of the converter (mixer). Stray capacitance from the open end of the gimmick to the converter circuitry provides positive feedback (regeneration) which is accompanied by reduced IF bandwidth. In turn, characteristic signal distortion occurs which chiefly consists of excessive accentuation of the treble, midrange, or bass tones, depending on the precise setting of the tuning dial.

Observe also that as the amount of feedback is increased, the IF section

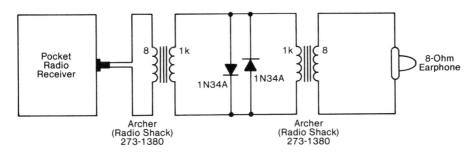

Note: This is a further familiarization experiment to demonstrate "the sound of clipping." It is similar to the preceding arrangement except that it uses two diodes shunted back to back across the 1k windings of the audio transformers. The clipping level is approximately ±0.2 peak volt. Observe that the intelligibility of the clipped speech signal is better than for either class-B distortion, or for class-AB distortion at moderate volume levels. However, background noise and spurious signal sound that tend to "follow" the modulation are more noticeable. Intelligibility of speech deteriorates at higher volume levels due to greater severity of clipping by the configuration.

Figure 4-3

A further familiarization experiment to illustrate "the sound of clipping."

will start to oscillate more or less; oscillation takes place at the IF peak frequency, in the vicinity of 455 kHz. The oscillatory voltage that is generated heterodynes with the incoming signal, with the result that a *"birdie"* is produced as the tuning dial is varied. Note also that in the event of strong IF oscillation, the incoming signal will be "wiped out," and the IF section will be dead for all practical purposes. (A comparatively high DC voltage will appear at the output of the detector diode.)

A supplementary familiarization experiment is shown in Figure 4-5 to acquaint the troubleshooter with the *"sound of audio instability."* This is also a positive-feedback malfunction; however, the fault is located in the audio section of the receiver. The signal distortion may be similar to "insufficient bandwidth," in some cases; however, there will not be a "birdie" in the sound as the tuning dial is varied. At higher levels of feedback, an audio tone or motorboating sound output occurs. At still higher feedback levels, the audio signal is completely wiped out, and only the spurious sound output is audible.

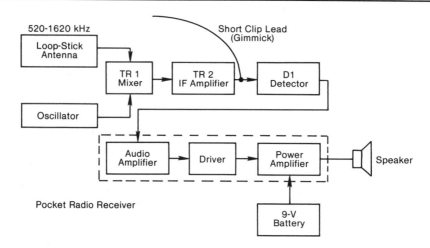

Note: This is another familiarization experiment to demonstrate "the sound of insufficient bandwidth." It uses a short clip lead "gimmick" from the input of the detector diode back to the vicinity of the mixer. The gimmick functions as a Q multiplier and reduces the IF bandwidth. If the gimmick is placed too close to the mixer section, the IF section will break into oscillation and squeal. Minimum bandwidth is provided just short of the point of oscillation. Observe that as a station is tuned in, "the sound of insufficient bandwidth" changes progressively from abnormal treble accentuation, through abnormal midrange accentuation, to abnormal bass accentuation. Musical passages are frequency-distorted, and speech may not be understandable.

Figure 4-4

An experimental arrangement to illustrate "the sound of insufficient bandwidth."

Figure 4-5

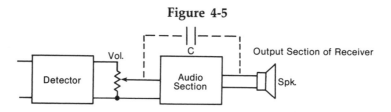

Note: This is a supplementary familiarization experiment to demonstrate "the sound of audio instability." It uses a clip lead and capaci-

Figure 4-5 (continued)

tor for feedback from the speaker to the volume control. If a Radio Shack 12-714 pocket radio is used, for example, C may have a value of 0.0004 μF. Observe that as the volume control is advanced (feedback is increased), the signal begins to sound scratchy. However, there is no "birdie" in the sound. Then, as the volume control is advanced farther, a low-frequency audio oscillation is produced which tends to mask the station signal. Beyond this point, the audio oscillation becomes dominant and wipes out the station signal.

Figure 4-5

A supplementary familiarization experiment to illustrate "the sound of audio instability."

USEFUL OLD AND NEW QUICK CHECKS

It is helpful to briefly recap some standard quick checks, along with informative follow-up tests for AM radio receivers.

1. If you are checking a battery radio, place your ear near the speaker; turn the power switch on and off. You will hear a click if the audio section is operative, if the speaker is workable, and if the earphone jack is not defective. If you do not hear a click, follow up by plugging in an earphone and repeating the click test.

2. In case the receiver passes the click test, turn up the volume control and listen for any noise (hissing) output. A reasonable amount of output noise throws suspicion on the receiver input circuitry (the local oscillator may have dropped out). Very low noise output directs attention to the detector circuit or the audio driver stage.

3. Follow up a very-low-noise condition with an *amplifier substitution test.* Locate the detector (it will be a diode), and feed the detector output into a mini-amplifier with a built-in speaker, such as the Archer (Radio Shack) 277-1008. You will now hear a normal noise level. If the signal sections are workable, you will also hear stations as the receiver tuning control is turned. (The detector diode is more likely to be a lN64, instead of a 1N34A.)

4. If you do not hear station signals in the foregoing test, follow up by connecting a 1N34A diode in series with the mini-amplifier input lead, and feeding the detector input signal into the 1N34A diode. If you now hear station signals as the receiver tuning control is turned, the detector diode in the receiver is defective. (See Figure 4-4.)

5. On the other hand, if you do not receive station signals using the substitute detector and amplifier, the trouble will be found in the RF (converter), IF, or AGC sections. Follow-up troubleshooting requires signal-section identification tests, as explained subsequently. (See Figure 4-8.)

6. There is still one more standard quick check that you will usually wish to make at this point. Lack of station-signal reception is sometimes caused by local-oscillator drop-out. In such a case, reception will be restored if the RF output from a signal generator is coupled into the converter circuit, or the antenna coil. (The generator must be tuned to the appropriate oscillator frequency.)

   Observe that there is normally considerable reradiation from the front end of an AM radio broadcast receiver. If you place a pair of normally operating receivers near each other and tune in a station on one receiver, you will hear "birdies" from the other receiver as it is tuned through its range (a particularly strong "birdie" will be heard 455 kHz from the station signal tuned in on the other receiver). *This is sometimes a useful quick check to determine whether the local oscillator has dropped out or whether the trouble is elsewhere.* Note that the local oscillator in an AM broadcast radio receiver operates "on the high side." If the tuning dial is set to 600 kHz, for example, the local oscillator will be operating at 600 + 455 kHz, or 1055 kHz. Observe that the incoming station-signal information is modulated on the local-oscillator "carrier," and that this information will be reradiated as translated 1055-kHz information to a nearby receiver. Harmonics and intermodulation products are also generated; they become more evident as the two receivers are placed nearer each other.

7. As noted in Chapter 3, localization time can often be saved by making temperature measurements of devices and components to see if the measured values are "in the ballpark." Measured values are most meaningful when made on a comparative basis with a unit that is in normal operating condition.

   Caution: Troubleshooters sometimes vary the V_{cc} voltage to observe its effect on the trouble symptom. It may be found that the receiver resumes near-normal operation when the V_{cc} voltage is increased. However, it is poor practice to "repair" a receiver by increasing its rated V_{cc} voltage. Even if transistors or integrated circuits do not burn out, their normal operating temperatures will be exceeded, and their life expectancy will be shortened.

Typical "Ballpark" Temperatures for a Pocket Radio Receiver:

   Ambient: 21°C

   Small-signal transistors: 24° to 26°C

Resistors: up to 26°C

Electrolytic capacitors: 23°C

Chassis: 22°C

(Note that the heat sink for a power-type transistor typically operates at 44°C.)

TUNED IF SIGNAL TRACER

The high-frequency signal tracer shown in Figure 4-6 is not applicable for checking the signal output from the converter (mixer) in a receiver because the signal level is low and may be subnormally low in the event of front-end

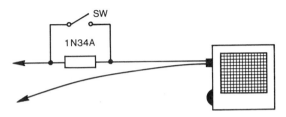

Note: A mini-amp/speaker functions as a "patch" section for quick checking the audio and speaker circuitry in a radio receiver. When a diode is included in the test lead, the tester also functions as a "patch" for the detector in the radio receiver.

Technical Note: After the malfunctioning section in a radio receiver has been localized, DC voltage and resistance measurements are ordinarily made to close in on the defective device or component. As explained previously, the greatest off-voltage value that is found (on a comparative test basis) often occurs at the terminals of the defective part—although there are exceptions, as in direct-coupled circuitry. It should be noted in this regard that if the fault is associated with an off-resistance value, the greatest off-resistance measurement will always occur at the terminals of the defective part. In the case of open capacitors, neither DC voltage nor resistance measurements are informative. Signal tracing (or signal substitution) is generally the most practical approach in this situation.

Figure 4-6

A useful two-in-one "patcher" for preliminary troubleshooting of radio receivers.

malfunction. However, the simple arrangement shown in Figure 4-7 provides ample sensitivity and serves satisfactorily as a tuned-IF signal tracer. Its operating frequency is 455 kHz and it can be used only in intermediate-frequency circuitry.

AM RADIO RECEIVER LAYOUTS

When troubleshooting without service data, it is helpful to keep the basic radio receiver layouts in mind. This knowledge assists in circuit tracing and identification of sections and stages. Widely used or standard broadcast AM radio arrangements are exemplified in Chart 4-1. Observe that there is a trend to use transistors in the converter and IF sections, followed by an integrated-circuit audio section.

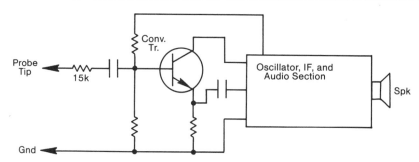

Note: This is a typical arrangement for conversion of a miniature AM radio into an IF signal tracer. The loopstick antenna is disconnected from the input coupling capacitor of the converter transistor, and a 15-k resistor is connected in its stead. The 15-k resistor becomes the probe tip for the signal tracer. Set the oscillator capacitor to minimum capacitance to avoid possible interference with the signal from the receiver under test. Mount the 15-k resistor so that it protrudes outside the case of the miniature radio to serve as a probe. In application, signal output will be heard from the speaker whenever the probe is touched to a point in the receiver under test that has a 455-kHz signal present. The Realistic (Radio Shack) AM pocket radio No. 12-201 will be found useful and convenient. Merely unsolder the loopstick antenna connection from the converter coupling capacitor, and solder a 15k probe resistor in place of the antenna connection.

Figure 4-7

Arrangement of a miniature AM radio for use as an IF signal tracer.

Chart 4-1

Widely Used AM Radio Transistor Arrangements

These are some basic AM radio receiver arrangements to keep in mind when troubleshooting without service data.

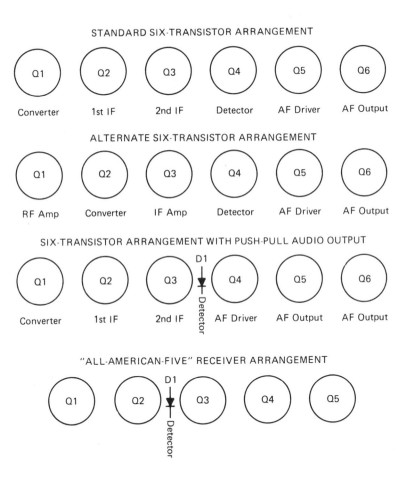

STANDARD SIX-TRANSISTOR ARRANGEMENT

Q1 — Converter
Q2 — 1st IF
Q3 — 2nd IF
Q4 — Detector
Q5 — AF Driver
Q6 — AF Output

ALTERNATE SIX-TRANSISTOR ARRANGEMENT

Q1 — RF Amp
Q2 — Converter
Q3 — IF Amp
Q4 — Detector
Q5 — AF Driver
Q6 — AF Output

SIX-TRANSISTOR ARRANGEMENT WITH PUSH-PULL AUDIO OUTPUT

Q1 — Converter
Q2 — 1st IF
Q3 — 2nd IF
D1 — Detector
Q4 — AF Driver
Q5 — AF Output
Q6 — AF Output

"ALL-AMERICAN-FIVE" RECEIVER ARRANGEMENT

Q1
Q2
D1 — Detector
Q3
Q4
Q5

Radio circuit boards sometimes have parts numbers marked beside the devices and components, starting with "1" for input parts, and increasing sequentially through the network, with the highest parts numbers marked beside the output devices and components.

The standard six-transistor arrangement and the alternate six-transistor arrange-

Chart 4-1 (continued)

ment employ transistor detectors. A transistor detector is essentially a demodulator and audio amplifier.

TWO-TRANSISTOR/INTEGRATED-CIRCUIT ARRANGEMENT

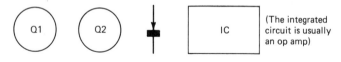

(The integrated circuit is usually an op amp)

PRELIMINARY ALIGNMENT PROCEDURE

After a transformer, antenna loopstick, or capacitor has been replaced in the RF or IF section of a radio receiver, alignment may become necessary for normal reception. Sometimes, alignment touch-up becomes desirable after an RF or IF transistor has been replaced. Preliminary alignment procedure provides satisfactory receiver operation, although more precise adjustments require the use of a signal generator and AC voltmeter.

Small AM broadcast radio receivers typically provide two trimmer capacitors and four ferrite cores for alignment, as shown in Figure 4-8. The IF transformers are designed to operate at 455 kHz. The tuning capacitors provide a signal frequency range of approximately 600 kHz to 1600 kHz. Each tuning capacitor is provided with a trimmer capacitor. One of the tuning capacitors varies the resonant frequency of the antenna loopstick, and the other tuning capacitor varies the resonant frequency of the local oscillator. (The tuning capacitors are ganged.)

Observe that the *positions of the ferrite cores in the antenna loopstick and oscillator coils are adjusted with respect to reception of station signals in the vicinity of*

Figure 4-8

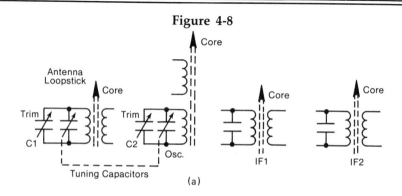

(a)

Figure 4-8 (continued)

Note: The alignment adjustments for a typical small radio receiver are depicted above. If alignment should become necessary, preliminary alignment procedure is as follows:

1. Turn the tuning control to receive a station signal near the low-frequency end of the dial (about 600 kHz). The dial normally indicates the correct frequency for the station. If it does not do so, then adjust the core of the oscillator transformer as required to obtain a correct indication on the tuning dial.

2. Next, slide the coil of the antenna loopstick into or out of the core as may be necessary to obtain maximum sound output from the receiver.

3. Then turn the tuning control to the other end of the dial to receive a station signal (about 1600 kHz). The dial normally indicates the correct frequency for the station. If it does not do so, then adjust the trimmer capacitor C2 on the oscillator tuning capacitor to provide a correct indication on the tuning dial.

4. Repeat the first and third steps of the procedure to obtain maximum sound output.

5. Next, turn the tuning control to again receive the station signal at about 600 kHz on the dial. Readjust the coil in the antenna loopstick as may be required for maximum sound output.

6. Now, turn the tuning control to receive a station signal at about 1400 kHz on the dial. Adjust trimmer capacitor C1 on the antenna loopstick tuning capacitor to obtain maximum sound output.

7. Repeat steps 5 and 6 to obtain maximum sound output. Then melt a drop of wax at the end of the coil on the antenna loopstick to secure it in final position.

8. Again turn the tuning control to receive the station signal at about 600 kHz on the dial. Adjust the core in IF transformer IF1 for maximum sound output. Also adjust the core in IF transformer IF2 for maximum sound output.

9. Repeat the core adjustments in IF1 and IF2 for maximum sound output. The preliminary alignment procedure is then completed.

Figure 4-8 (continued)

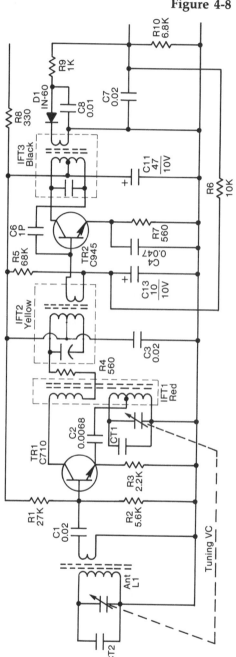

Generator Setting	Dial Setting	Adjust	For
1. 455 kHz	60	IFT 2(yellow)	Max out
2. 455 kHz	60	IFT 3(black)	Max out
3. Repeat 1 and 2			
4. 600 kHz	60	IFT 1 (OSC Coil red)	Max out
5. 1600 kHz	160	Trimmer CT-1	Max out
6. Repeat 4 and 5			
7. 600 kHz	60	Antenna Coil	Max out
8. 1400 kHz	between 160 and 100	Trimmer CT-2	Max out
9. Repeat 7 and 8			

(b)

Note: This is a diagram of the tuned circuitry provided in the ArcherKit® (Radio Shack) AM radio kit, Cat. No. 28-4029. The DVM may be a Micronta (Radio Shack) digital multimeter Cat. No. 22-195.

Figure 4-8

Typical alignment procedure: (a) preliminary alignment procedure for a small AM broadcast radio receiver; (b) tuned circuitry and alignment procedure with test instruments for a popular pocket radio receiver.

600 kHz. On the other hand, the trimmer capacitors are adjusted with respect to reception of station signals in the vicinity of 1600 kHz. [These alignment adjustments provide optimum correlation ("tracking") of the antenna loopstick and oscillator resonant frequencies.]

ALIGNMENT WITH TEST EQUIPMENT

Consider next how the alignment procedure explained in Figure 4-8 is more precisely accomplished with the aid of test equipment. An AM RF signal generator and DVM are employed. To inject the test signal from the generator into the radio receiver circuitry, a radiation loop consisting of several turns of wire about 60 cm in diameter may be connected to the generator output cable and placed in proximity to the radio receiver. Or, a few turns of wire may be wrapped directly around the PC board.

In either case, use only as much signal strength as required to obtain ample indication on the DVM. This ensures that the radio receiver is operating at maximum gain (minimum AGC voltage). In turn, the alignment indications become more precise (sharper), and the troubleshooter will be certain that the receiver circuits have been aligned for optimum response to weak signals. The DVM is connected across the speaker terminals, and is operated on its low-range AC function. The volume control is turned to maximum.

The tuned circuitry depicted in Figure 4-8(a) is shown in more detail in Figure 4-8(b), and the alignment procedure is listed in nine steps on the accompanying chart. The antenna coil is initially positioned as explained in Figure 4-8(a). After Step 9 has been completed in Figure 4-8(b), a drop of wax is melted with a soldering gun, and the antenna coil is secured in position on the ferrite core with the wax.

AUTODYNE CIRCUIT TROUBLESHOOTING

Most small AM radio receivers use autodyne circuitry between the antenna and the IF section, as exemplified in Figure 4-9. An autodyne circuit employs a single transistor to function both as a local oscillator and a heterodyne converter. Heterodyne conversion requires nonlinear operation of the transistor (such as class-B or class-C operation). However, *the oscillator section will not be self-*

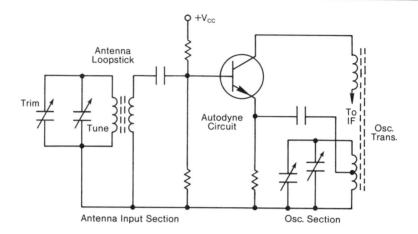

Antenna Input Section   Osc. Section

Technical Note: The autodyne circuit functions both as a local oscillator and a heterodyne converter. In order for the local oscillator to be self-starting, forward-bias voltage must be applied to the base and emitter terminals. After oscillation starts, rectifier action in the base-emitter circuit reverse-biases the transistor more or less, and base current flows in pulses.

As a practical example of normal operation, the base-emitter bias is 0.5 volt (transistor is cut off on the average). If the oscillator stops, the base-emitter voltage rises to 0.65 volt (transistor is forward-biased). Unless the oscillator is operating, there is no IF signal output.

Note: The chief disadvantage of the autodyne circuit (particularly at high frequencies) is the tendency of the oscillator to "pull" with respect to the frequency of the incoming signal, with resultant shift in effective alignment. Accordingly, high-performance receivers such as amateur communications receivers typically employ separate oscillator and mixer stages for reduction of coupling between the incoming signal circuit and the oscillatory circuit.

Figure 4-9

Example of autodyne circuit operation.

starting unless the transistor is biased into its class-A region. These conflicting requirements are met by means of a combination of voltage-divider bias and self (signal-developed) bias.

With reference to Figure 4-9, the base voltage divider normally supplies 0.65 V of base-emitter forward bias. In turn, unless there is a defect in the oscillator section, oscillation will start. As soon as the oscillator generates an AC voltage in the emitter-collector circuit, the transistor supplies its own input and this positive feedback drives the transistor into saturation on one excursion, and into cutoff on the other excursion. This is just another way of saying that the transistor then operates nonlinearly and that oscillator signal rectification takes place in the base-emitter circuit.

The practical result from the troubleshooter's viewpoint is that reverse-bias voltage builds up on the base coupling capacitor and that, on the average,

Figure 4-10

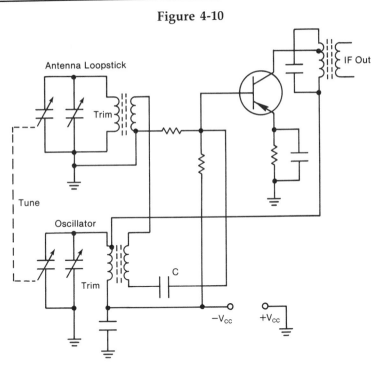

Note: This type of autodyne circuit is less common than that shown in Figure 4-11. However, it will be encountered on occasion. Observe that the chief difference between this arrangement and the former configuration is that the oscillator section operates in the base branch of this network. Feedback takes place from collector to base in this design, instead of from collector to emitter as in the more common

Figure 4-10 (continued)

arrangement. However, circuit action is essentially the same in both configurations. Thus, when the power is first applied, the transistor is biased for class-A operation, but signal-developed reverse bias quickly builds up on C. In turn, the transistor is normally biased (on the average) into the class-B region. As in the preceding example, base current normally flows in sharp high pulses.

Figure 4-10

Another example of autodyne circuitry.

the transistor operates in class B. However, *the transistor is not actually cut off because substantial base current flows in high narrow pulses* on the positive peaks of the AC base signal input. Observe that a leaky coupling capacitor or an off-value bias resistor can cause a "dead oscillator" trouble symptom.

When troubleshooting without service data, it is helpful to keep basic circuitry variations in mind so that the standard configurations will be recognized. Figure 4-10 shows another version of the autodyne circuit that will be encountered in various older types of small AM broadcast receivers. Similar to the arrangement shown in Figure 4-9, *this configuration also uses a combination of voltage-divider and signal-developed base bias.* However, this arrangement operates with the tuned oscillator transformer in the base branch.

Observe that the local oscillator appears to be "dead" in a receiver that has normal noise output but no audio output. This may be the case if the coupling capacitor to the oscillator section is open, for example. On the other hand, the oscillator is not actually "dead" if the connection to the trimmer capacitor becomes defective—the oscillator merely operates off-frequency in this situation. Note that if the oscillator is functionally "dead," class-A bias voltage will be measured, but if the oscillator is only operating off-frequency, class-B bias voltage will be measured.

Puzzling "weak-reception," distortion, and "birdie" problems can sometimes be solved by an impedance check at the battery clip terminals, as explained in Chart 4-2.

FM/AM RADIO RECEIVER TROUBLESHOOTING

A few pocket radio receivers provide FM reception only, but most small FM receivers also are combined with AM reception. It is easier to "size up" a receiver that provides FM reception only because there are fewer components, transistors, and/or integrated circuits to contend with. Note that in the case of receivers with several functions and complex circuitry, comparison tests may be the only practical procedure when troubleshooting without service data.

In the case of a small combination FM/AM receiver with normal AM

Chart 4-2

Impedance Check at Battery Clip Terminals

"Tough-dog" troubleshooting problems can be caused by open capacitors (and capacitors with a poor power factor) associated with the V_{cc} line.

Such defective capacitors can often be caught at the outset by making a comparison check of the impedance Z at the battery clip terminals.

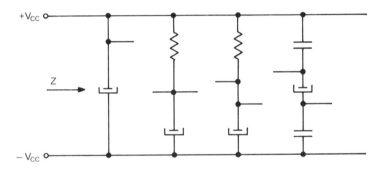

An impedance comparison check of a good receiver and a bad receiver is made with an audio generator, a DVM, and a resistor, as shown below.

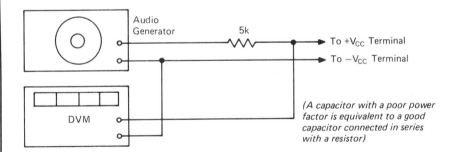

(A capacitor with a poor power factor is equivalent to a good capacitor connected in series with a resistor)

The comparison impedance check is made by disconnecting the battery from the receiver, and connecting the test leads of the checker to the V_{cc} terminals. Operate the audio generator at a low frequency, such as 60 Hz. Advance the output voltage from the generator to obtain an adequate AC-voltage indication on the DVM. Then, repeat the test on the other receiver, and compare the two readings.

Chart 4-2 (continued)

Example: A typical pocket radio receiver in normal operating condition produces a reading of 20 mV on the DVM when the foregoing impedance test is made. If the reading were significantly excessive, such as 100 mV, or 500 mV on a bad receiver, the troubleshooter would look for one or more capacitors in the V_{cc} circuit with defects such as poor power factor, open circuit, or loss of capacitance conditions.

A quick check of a suspected open capacitor (or loss of capacitance, or poor power factor) can be made by temporarily bridging a good capacitor across the terminals of the suspect capacitor. Then, if normal operation resumes, the troubleshooter has localized the fault.

reception but with trouble symptoms in FM reception, the basic foregoing quick checks can be repeated on the FM section and the results can be compared. However, note the following technical points:

1. Most FM/AM radios employ more transistors on the FM function than on the AM function. In turn, the normal current drain with the volume control turned down is somewhat higher. For example, a radio that draws 9 mA on its AM function draws 11.5 mA on its FM function. (Connect a milliammeter in series with the battery.)
2. To check for a dead local oscillator, place the defective FM receiver near a normally operating FM receiver. In this case, the technician is not listening for "squeals"—he or she is listening for "mutes."

   Example: When the operating FM receiver is tuned to a station at 107 MHz, and the tuning dial of the dead receiver is varied from 88 to 98 MHz, a "dead silence" at 96.3 MHz (107–10.7 MHz) indicates that the local oscillator in the dead receiver is workable.

CAPTURE EFFECT

The foregoing quick check of local-oscillator operation in an FM receiver is based on the *capture effect*. In other words, when two FM carriers are present at the same frequency, the weaker signal will be completely suppressed and the stronger signal will capture the FM detector. In this quick check, the weaker FM signal is the station to which the operative FM receiver is tuned. The local-oscillator radiation from the supposedly dead FM receiver is a stronger FM signal because the receivers are side by side. Inasmuch as the radiated local-oscillator signal is unmodulated, capture results in the silencing of the operative receiver.

You will observe that when the supposedly dead FM receiver is moved away from the operative receiver, the capture effect ceases and the operative receiver reproduces the station signal to which it is tuned.

Then, as you move the supposedly dead FM receiver still farther away from the operative receiver, the capture effect again occurs, and the operative receiver is again silenced. This test result occurs because of *standing waves* in the room. The local-oscillator radiation in this example has a frequency of 107 MHz, and the peaks of the standing waves are spaced 4.6 feet apart.

THE "SOUND OF FULL RECTIFICATION"

It is helpful for the troubleshooter to know *"the sound of full rectification"* in signal-tracing procedures. A simple demonstration experiment for familiarization with "the sound of full rectification" is shown in Figure 4-11. The audio output signal from an ordinary pocket radio receiver is passed through a full-wave bridge rectifier. Consequently, the negative excursions of the audio waveform are converted into positive excursions, and the bridge output is "all positive." Observe that the full-rectified signal has a harsh and "crackly" timbre that varies considerably with the audio content. However, speech in general remains intelligible.

IC FUNCTIONAL OVERVIEW

An AM radio receiver system with all of the active portions of a conventional AM receiver contained on a single IC chip is typified in Figure 4-12. Only the resonant circuitry has to be provided externally. In this example, the IC includes the RF converter, IF amplifier, detector, AVC circuit, built-in regulator

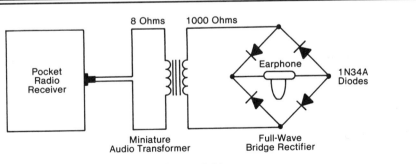

Figure 4-11

Familiarization experiment for demonstration of "the sound of full rectification."

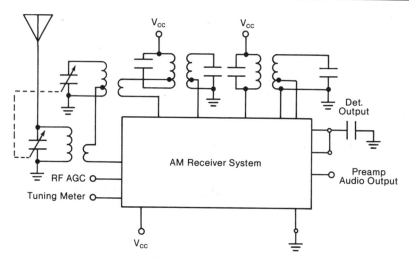

Note: Although a transistor is a single-function device, an integrated circuit is a multifunction device. Observe that the IC in this example functions as an autodyne converter, IF amplifier, second detector, audio preamplifier, and AGC circuit. A built-in voltage regulator is also included. In general, IC testers similar to transistor testers are not available. Therefore, the troubleshooter must test the IC on the basis of signal tracing. (Is signal voltage present or absent at a particular IC pin?) If signal voltage is absent, the reason may be found in the external circuitry, or in the IC. The external circuitry is checked in the same manner as for transistorized receivers. If no fault is found in the external circuitry, it is logical to conclude that the IC is defective.

Figure 4-12

Arrangement of an integrated-circuit AM radio receiver system.

zener diode, and an audio preamp stage. Note that in some AM radio receiver system ICs, an RF amplifier is included, with the audio preamp or tuning meter omitted. Basic functional parameters are:

1) *Sensitivity.* The rated overall receiver sensitivity is based on a specified RF coil and IF coil arrangement, usually at 1 MHz, with 30 percent amplitude modulation, an audio frequency of 400 Hz, and for a specified output voltage level. Thus, a typical sensitivity for 10 mV output would be 3 microvolts.

2) *Signal to noise ratio.* The rated signal-to-noise ratio is measured under the same conditions as above. A typical rating is 45 dB.

3) *Maximum power dissipation.* The rated maximum power dissipation is generally measured at room temperature. A typical AM radio receiver system IC is rated to dissipate 600 mW.

This type of IC is used in miniature and subminiature AM radio receivers of the broadcast and weather types and in various other consumer products.

A Dolby noise-reduction processor for single audio channels contained on an IC chip is depicted in Figure 4-13. It provides Dolby B-type noise-reduction features, and is found in various audio amplifiers and the audio channels of

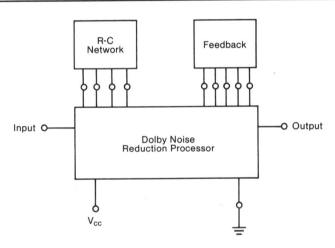

Note: A Dolby noise reduction processor integrated circuit is a multi-function device used to improve the reproduction of tape recording sound. In the Dolby system, low sound levels are boosted prior to recording and are then reduced on playback for the purpose of eliminating tape hiss. The Dolby system is also used in FM broadcasting (Dolby FM). Dolby A is a professional design that employs four separate frequency channels. Dolby B is used in home entertainment equipment and has only one channel. It provides 10 dB of noise reduction at frequencies above 5 kHz. ("Dolby" is the trademark of Dolby Laboratories, Inc.)

Figure 4-13

Arrangement of an integrated-circuit Dolby noise-reduction processor.

some VCRs. In addition to an internal power-supply regulator, the IC package comprises a series of amplifiers associated with external RC networks. One of these networks, containing five capacitors and three resistors, is connected to four external leads. The second RC network forms a feedback path and consists of three resistors and three capacitors plus an internal rectifier circuit. The basic functional parameters are:

1. *Distortion.* The maximum rated distortion contributed by the IC is specified as 0.05 percent for a 1-kHz, 0-dB level input and as 0.1 percent for a 10-kHz, 10-dB level input.

2. *Signal handling.* This rating describes the dynamic range of the signal for 0.3 percent distortion at 1 kHz (14 dB is a typical value).

3. *Signal-to-noise ratio.* Rated for the encode mode of operation. (A typical value is 70 dB.) Rated also for the decode mode of operation. (A typical value is 80 dB.)

4. *Input resistance.* A typical rating is 65 kilohms.

5. *Output resistance.* A typical value is 80 ohms.

Next, an integrated-circuit FM/IF system is shown in Figure 4-14. This IC contains a three-stage FM-IF amplifier/limiter arrangement with level detectors

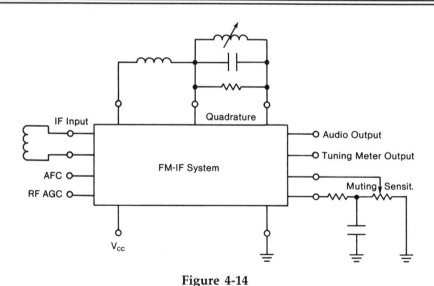

Figure 4-14

Arrangement of an integrated-circuit FM-IF system.

for each stage and a balanced quadrature FM detector. As shown in the diagram, the quadrature-tuned circuit requires two external coils, a capacitor, and a resistor (with a single tuning adjustment). Most ICs of this type also include an audio amplifier and a driving circuit for tuning-meter output, automatic frequency control (AFC) for the FM tuner, and AVC for the RF amplifier stage.

The muting sensitivity control in Figure 4-14 is optional and provides automatic muting (squelching). An internal power supply regulator enables the use of any supply voltage between 8 and 18 volts. The amount of distortion produced by this IC depends chiefly on the phase linearity of the quadrature-detector coil. Basic functional parameters are:

1) *Input limiting voltage.* The rated voltage applied at the input terminal that will cause limiting of the FM-IF signal (10 microvolts is a typical value).

2) *AM rejection.* The rated rejection of amplitude modulation at an input voltage of 100 mV, with 30 percent amplitude modulation at 400 Hz (55 dB is a typical rated value).

3) *Recovered audio voltage.* Rated amplitude of the audio output signal when the input signal reaches its limiting level with 400 Hz audio, FM modulated at 25 kHz. (Typical ratings are from 300 to 500 mV.)

4) *Total harmonic distortion.* Rated audio distortion at all frequencies with input conditions as above. A typical rating is 0.5 percent with a single-tuned input circuit, or 0.1 percent with a double-tuned input circuit.

5) *Signal-to-noise ratio.* Measured with a deviation of ±75 kHz. (A typical rating is 65 dB.)

This design of IC is used in all types of FM receivers, and is intended for 10.7-MHz IF operation. To provide sufficient FM-IF selectivity, many designers of FM receivers employ a narrow-band filter before the IF input stage. Note that because of the very high gain of this IC, PC board layout and the location of bypass and decoupling capacitors become critical. Routing of grounds is also critical.

Next, an integrated circuit stereo multiplex decoder is exemplified in Figure 4-15. This IC accepts its input signal from the FM detector and applies it simultaneously to a 19-kHz and a 38-kHz synchronous detector. A 76-kHz local oscillator is part of the IC, and is controlled by the external circuits shown in the diagram. This local oscillator generates a signal which is counted down to 38 kHz and counted down to two 19-kHz signals in phase quadrature. These signals are used in the two synchronous detectors.

The incoming FM pilot signal is compared against a preset reference level in one of the detectors. When it exceeds a threshold value, the pilot signal sets a Schmitt trigger circuit that energizes the stereo indicator light. This enables the 38-kHz synchronous detector and automatically switches the circuitry from monaural to stereo operation. Internal audio preamps develop the L and R

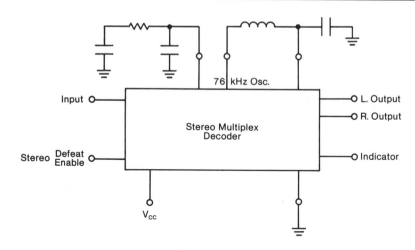

Figure 4-15

Arrangement of an integrated-circuit stereo-multiplex decoder.

output signals. External control of stereo "defeat" or "enable" is provided by a separate terminal on the IC. Basic functional parameters are:

1. *Input impedance.* A typical rating is 50 kilohms.

2. *Channel separation.* Expresses the relative interference between the L and R channels. (A typical rating is 0.3 dB.)

3. *Channel balance.* This rating denotes (in the monaural mode) the difference between the L and R channels. (A 0.3-dB rating is typical.)

4. *Monaural gain.* Denotes the rated amount of monaural amplification. (A typical value is 6 dB.)

5. *Stereo/monaural gain ratio.* Denotes the difference in gain between monaural and stereo operation. (A ±0.3-dB rating is typical.)

6. *Capture range.* The rated deviation from the 76-kHz center frequency over which the IC will operate. (A rating of ±10 percent is typical.)

7. *Distortion.* Rated amount of second harmonic distortion. (A value of 0.2 percent is typical, with less than 0.2 percent at higher harmonic frequencies.)

8. *19-kHz rejection.* Rated reduction in 19-kHz frequency component in the audio output signal (35 dB is typical).

9. *38-kHz rejection.* Rated reduction in 38-kHz frequency component in the audio output signal (48 dB is typical).

This stereo multiplex decoder IC is used in many FM-stereo systems. It requires one low-inductance tuning coil.

SWEEP ALIGNMENT WITH THE OSCILLOSCOPE

Some troubleshooters prefer to align FM circuits with a sweep-frequency generator and oscilloscope, as shown in Figure 4-16. Observe that the bandwidth of the IF response curve is 200 kHz, and normally extends from 10.6 MHz to 10.8 MHz. The FM demodulator response curve has a peak-to-peak excursion (bandwidth) of 200 kHz, from 10.6 MHz to 10.8 MHz, with a center frequency of 10.7

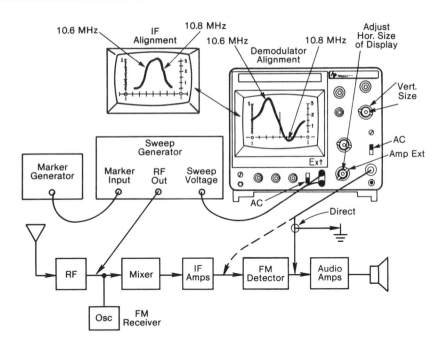

Courtesy of B&K Precision, Div. of Dynascan Corp.

Figure 4-16

FM receiver sweep-alignment setup.

MHz. A marker generator is used with the sweep generator to display a "pip" on the response curve at the frequency to which the marker generator is set.

A frequency marker (pip) identifies any chosen frequency point along the response curve. Note that the IF response curve is displayed when the oscilloscope is connected at the input to the FM detector. (A demodulated waveform is displayed because the last IF stage operates as a limiter.) The FM demodulator response curve is displayed when the oscilloscope is connected at the output of the FM detector.

The scope and sweep-frequency generator are troubleshooting instruments, as well as alignment instruments. Thus, if a frequency response curve cannot be brought within acceptable tolerances, it follows that there is a malfunction in the associated circuitry. If the troubleshooter knows "how to read sweep-frequency waveforms," he or she can rapidly obtain helpful clues concerning the type of malfunction, if not the specific component or device fault. It may be noted that the output response curve from one stage is normally multiplied by the response curve of the second stage, on the basis of individual stage responses. However, if there is spurious feedback (regeneration) present, the overall response of the two stages will be quite different from the simple product of the individual stage responses.

5

ADDITIONAL RADIO TROUBLESHOOTING TECHNIQUES

PC Through-Board Locater • FM Front-End Familiarization • Scavenging Tricks of the Trade • Miscellaneous Quick-Check Examples • FM/AM Radio-Phono Arrangements • Comparative DC Voltages with and without Signal • Stereo FM Pocket Radio • Avoidance of High-Frequency Circuit Disturbance • Comparative DC Voltage and Resistance Checks • Comparative Resistance Checks • Automatic Frequency Control • Basic Plan of FM/AM Integrated-Circuit Receiver • Front End with Separate Oscillator and Mixer Transistors • Is the FM Converter Working? • PC Board Layout Principles • "Detector Spotter" Operation of AM Radio Receiver • Converter Voltages in Oscillatory and Non-Oscillatory States • All Voltages Negative • Oscillator Signal Substitution • Second-Harmonic Generation by FM Detector • Automobile Radios • IC Functional Overview

PC THROUGH-BOARD LOCATER

When troubleshooting radio receivers without service data, various parts such as transistors are readily apparent on the component side of the PC board, but the corresponding transistor terminals are often difficult to locate on the solder side of the board. If the PC board is translucent, a strong light can be placed behind the board, and the shadows of the devices and components then become visible over their corresponding terminals on the solder side of the board.

However, not all PC boards are sufficiently translucent to identify terminals in this manner. Accordingly, other methods must be used to "buzz out" terminals and circuitry. There are several tricks of the trade that are useful in this procedure. One of the simplest locaters is a conventional outside caliper. The PC board is placed between the legs of the calipers as depicted in Figure 5-1. In turn, when the top leg of the caliper is lowered over a lead or eyelet at any point on the board, the upper leg of the caliper then rests at the corresponding solder pad on the solder side of the board.

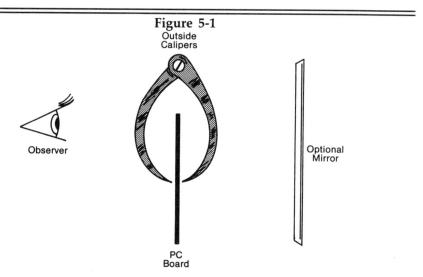

Figure 5-1
Outside
Calipers

Observer

Optional
Mirror

PC
Board

Note: The calipers serves to indicate corresponding pigtails or eyelets and solder pads from the component side of the PC board to the solder side. An optional mirror can save considerable time and an-

Figure 5-1 (Continued)

noyance in moving the calipers through a succession of points on the component side of the board. In other words, the mirror enables the troubleshooter to look at both sides of the board without "flipping" the board over.

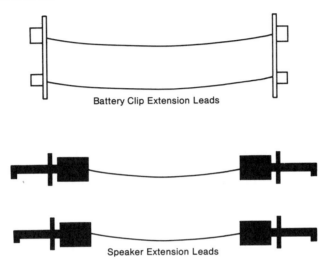

Battery Clip Extension Leads

Speaker Extension Leads

Note: When "sizing up" a PC board during preliminary troubleshooting procedures, the board often needs to be lifted or turned over or rotated. These changes in position are hampered by the comparatively short battery clip leads and speaker leads. However, the troubleshooter can easily make up a pair of battery clip extension leads, and use a pair of microtest leads to provide ample freedom in moving the PC board as required. (The leads to the speaker must be temporarily unsoldered for connection of the microtest extension leads.)

Figure 5-1

Handy troubleshooting items: (a) an outside calipers used as a PC through-board locater; (b) extension leads for battery and speaker.

An alternative method of through-board location is shown in Figure 5-2. Here, a pair of transparent grids are used to check the "longitude" and "latitude" of corresponding points on the component side and the solder side of the PC board. The transparent grids are made up by ruling horizontal and vertical lines on a pair of clear plastic sheets. When the upper left-hand corner of the PC

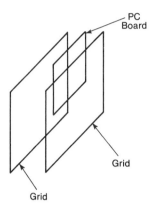

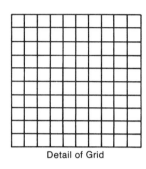

Detail of Grid

Note: This method of through-board location makes use of a pair of transparent ruled grids. To facilitate "counting off squares," the horizontal lines may be numbered 1, 2, 3, and so on, and the vertical lines may be lettered A, B, C, and so on. Generally, it is easiest to lay the PC board flat on the bench, component side up, and then to lay a grid over the board. The location of the desired point is noted. Then the PC board is turned over and the grid is again laid over the board. In turn, the selected point can be located on the solder side of the board. Or, the troubleshooter may choose to construct a jig for holding the PC board and a pair of grids vertically. This is a more practical arrangement when a sucession of points are to be located.

Figure 5-2

Alternative method of through-board location.

board is placed in line with the upper left-hand corners of the grids, corresponding points on opposite sides of the PC board become readily apparent.

FM FRONT-END FAMILIARIZATION

When the oscillator section is defective, it is helpful to recognize basic front-end arrangements. Nearly all pocket type FM radios and many table-model radios have an aperiodic RF input stage, as shown in Figure 5-3. Observe that the FM antenna circuit is untuned; it consists simply of a small coil shunted by a small fixed capacitor. Since the RF transistor is operated in the common-base mode, its input resistance is very low (on the order of 35 ohms). Consequently, the aperiodic input circuit has practically the same impedance over the entire FM band, and "looks" more like a resistor than a tuned circuit.

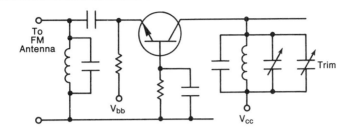

Note: This RF amplifier FM input arrangement is in very wide use for small broadcast FM and FM/AM radio receivers. The transistor operates in the CB mode and in class A. Coils are small and self-supporting with several turns of enameled wire. The emitter circuit is fixed-tuned and has very low impedance. However, the collector circuit is variably tuned with a (ganged) tuning capacitor. The collector coil is shunted by its tuning capacitor, by a small fixed capacitor, and by a trimmer capacitor. As a practical note, if the transistor is replaced, it is advisable to "touch up" the alignment of the trimmer capacitor for maximum sound output.

Figure 5-3

FM front end with aperiodic RF input circuit and tuned output circuit.

Observe in Figure 5-3 that the collector circuit of the RF transistor is tuned; it has a much higher impedance than the emitter circuit and functions to provide both selectivity and substantial voltage gain. Next, observe the typical FM autodyne converter configuration shown in Figure 5-4. It has a very low-impedance and broadly fixed-tuned input circuit, with a variably tuned collector circuit. A diode is included in the collector branch to facilitate heterodyne action. As noted previously for the RF section, if the transistor is replaced, it is advisable to "touch up" alignment of the trimmer capacitor. (See also Figure 5-5.)

SCAVENGING TRICKS OF THE TRADE

Replacement parts procurement can be a problem when troubleshooting radio receivers without service data. For example, transistors are seldom identified, although it is safe to assume that they are silicon types. Whether a PNP or NPN type of transistor is used in a circuit is evident from the polarity of the collector voltage (NPN transistors operate with positive collector voltage, and PNP transistors operate with negative collector voltage). Power ratings can be estimated from the type of circuit.

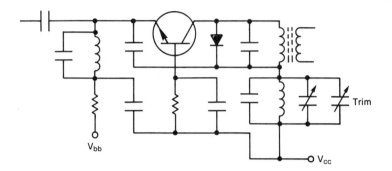

Note: This type of FM autodyne converter arrangement is in very wide use for small broadcast FM and FM/AM radio receivers. It is basically similar to the AM autodyne converter arrangement noted previously, except that the transistor operates in the CB mode, and the input circuit is essentially fixed-tuned. However, the collector circuit is variably tuned with a tuning capacitor, supplemented by a small fixed capacitor and a trimmer capacitor. Note that if the transistor is replaced, the troubleshooter should "touch up" the alignment of the trimmer capacitor.

Figure 5-4

Typical autodyne converter arrangement for FM front end.

Troubleshooters should "never throw away anything." For example, transistors scavenged from discarded radio receivers are often completely satisfactory as replacements in a receiver that is being repaired. Observe that a transistor scavenged from the IF section of any AM receiver will operate in the IF section of any other AM receiver. Similarly, a transistor scavenged from the IF section of any FM receiver will operate in the IF section of any other FM receiver.

Most transistors are soldered into the associated PC board. In turn, they must be desoldered without thermal damage. If the transistor leads are accessible on the component side of the board, soldering tweezers or equivalent heat sink devices can be used to avoid possible thermal damage. However, if the transistor leads are not accessible on the component side of the board, the only recourse is to desolder the leads as rapidly as possible. (In production, the leads are soldered to their pads rapidly to avoid thermal damage.) Special desoldering tips and desoldering irons may be used to speed up the procedure.

The Archer (Radio Shack) Semiconductor Reference Guide lists over 100,000 semiconductor substitutions, and also provides a large amount of tech-

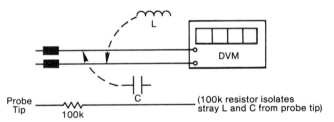

Probe Tip ——ᴡᴡ—— 100k —— C —— (100k resistor isolates stray L and C from probe tip)

Note: Any pair of test leads has more or less stray capacitance and inherent inductance. At high frequencies, the test leads "look like" a series or a parallel tuned circuit (resonant stub). In turn, the stub upsets high-frequency circuit action, and false DC voltage values are often indicated. However, if a 100k resistor is used as a probe tip, this resistance "isolates" the stray inductance and capacitance and permits measurement of correct DC voltage values. If both test leads are operated above ground potential, a 100k resistor should be used with each lead.

Figure 5-5

Stray capacitance and distributed inductance of long test leads causes high-frequency circuit disturbance.

nical information concerning semiconductor characteristics, packages, and typical applications. Operating considerations, general precautions, and simple tests for transistor condition are also included.

MISCELLANEOUS QUICK-CHECK EXAMPLES

Comparative current-drain tests are sometimes informative.

For example, a current meter can be connected in series with the V_{cc} battery in the bad receiver and in a similar good receiver. A typical good AM/FM pocket radio has a current demand of approximately 10 mA at low volume output, and about 25 mA at high volume output. If the bad receiver happens to draw 10 mA at low volume, but draws only 15 mA when the volume control is advanced to maximum (and has weak output), the troubleshooter would look for a malfunction in the audio section.

In another example, a typical good AM/FM portable radio has a current demand of approximately 11 mA at low volume output, and up to 30 mA at high volume output. If the bad receiver happens to draw 5 mA at low volume, and draws only 14 MA when the volume control is advanced to maximum (and has weak and/or distorted output), the troubleshooter would look for a malfunction in the RF or IF section.

Functional comparative current-drain tests are occasionally useful.

For example, an FM/AM portable radio normally has a current demand of 10 mA on its AM function at low volume, and draws 11 mA on its FM function at low volume. Then, if the bad receiver draws 10 mA on its AM function, and 20 mA on its FM function at low volume, the troubleshooter will conclude that the trouble will be found in a stage that is operative only on the FM function.

Relative oscillator-injection voltage measurements can finger a poor-sensitivity trouble symptom in some situations.

As an illustration, normal converter gain depends on the local-oscillator injection voltage. In other words, the converter transistor is both an oscillator and a heterodyne mixer. The incoming RF signal is mixed with the local-oscillator voltage in the converter transistor, and the resulting 455-kHz beat frequency is developed at the collector.

If the oscillator-injection voltage is weak, the result is a poor-sensitivity trouble symptom. (The cause of weak oscillator-injection voltage can be a leaky or open capacitor, for example.)

A relative oscillator-injection voltage measurement is a comparison quick-check made with respect to a similar good receiver. The check is made at the collector of the converter transistor, using a peak-reading probe and a DVM. For example, 0.6 volt is a typical ball-park value. A substantially lower value would indicate malfunction in the circuitry.

On the other hand, a good transistor from the AM IF section of a radio receiver is not likely to work satisfactorily in the FM IF section of the same receiver or another receiver. Any transistor has its own rated cutoff frequency (f_T), and a transistor with an f_T of 1 MHz, for example, will not work in a 10.7 MHz FM IF circuit. Similarly, a transistor with an f_T of 20 MHz will not work at 100 MHz in an FM autodyne circuit. Observe that a transistor with an f_T of 200 MHz, for example, will operate in an AM IF circuit, although its maximum available gain is likely to be less than the original transistor.

In general, audio transistors have the lowest f_T values and the highest gain values. Conversely, FM front-end transistors have the highest f_T values, and usually considerably lower gain values. Note that commercial replacement transistors with high f_T ratings are more expensive than comparable transistors with low f_T ratings. Insofar as diodes are concerned, any diode can be used in any section of an FM/AM receiver.

Capacitors can usually be scavenged without difficulty; the capacitance value and voltage rating are usually marked on electrolytic capacitors, and the capacitance value is usually marked on ceramic capacitors. Of course, this information greatly simplifies replacement procedures. Resistors also generally present little difficulty, inasmuch as they are almost always color-coded. As noted previously, the power rating of a resistor is approximately proportional to its size.

Scavenging of inductive components such as IF transformers or loopstick antennas is less straightforward. In most situations, the only practical procedure is to "try and see if it works." The same manufacturer is likely to use the same basic circuitry and the same components in various models of FM/AM receivers. In some cases, we find exactly the same PC boards used by the same manufacturer in pocket radios which have quite different outward appearances.

FM/AM RADIO-PHONO ARRANGEMENTS

Small table-model FM/AM radio-phono units employ the same radio circuitry that was described in the preceding chapter. In addition, a function switching position is provided for driving the push-pull output amplifier from a stereo phono cartridge. This type of equipment provides L and R speaker outputs on the phono function only. That is, only mono reproduction is provided by the speakers when the equipment is operated on its FM function.

An AC power supply is typically provided, with the power-supply transformer operated as part of the phono motor assembly. A common arrangement uses half-wave rectification filtered by a 1000 μF electrolytic capacitor. The push-pull output amplifier is ordinarily a pair of integrated circuits with several external resistors and capacitors. A ganged volume control operates on all three functions.

STEREO FM POCKET RADIO

In addition to AM-only pocket radios, we will find stereo FM-only pocket radios. This type of receiver operates with earphones, and cannot drive a speaker. The antenna is integral with the earphone cord. A 28-pin integrated circuit is employed with one transistor. Because the circuit board is more densely packed than other types of pocket radio receivers, the troubleshooter may not wish to cope with the circuitry unless a fault happens to be comparatively obvious. A considerable number of very small components are mounted on the circuit board. However, it is feasible to make comparison tests with a similar receiver in normal operating conditions.

AUTOMATIC FREQUENCY CONTROL

We will find automatic frequency control (AFC) circuitry in the more elaborate FM radio receivers. AFC functions compensate for oscillator drift thereby keeping the receiver tuned "on station." Thus, if the local oscillator frequency happens to drift slightly off its correct value (10.7 MHz above the station frequency), the AFC circuit reacts to bring the oscillator back to practically its normal operating frequency.

Note that this frequency-control action has definite limits, and in case the oscillator drifts excessively, the control action suddenly stops and the receiver will lose the station to which it is supposedly tuned. Figure 5-6 depicts the plan of an AFC system. Observe that a reverse-biased diode (varactor diode) is

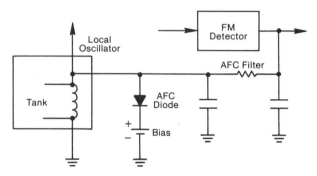

Note: A switch is generally provided to short out the AFC control voltage, if desired. This is helpful when tuning in comparatively weak stations, because the AFC action "pulls" the local-oscillator frequency appreciably before it loses control. In turn, the oscillator then jumps quickly to its normal frequency, and in doing so may "skip" over a comparatively weak station signal. In other words, a weak station should first be tuned in with the AFC switch turned off; after it is tuned in, the AFC switch may then be turned on.

Reproduced by special permission of Reston Publishing Company from Radio Theory and Servicing *by Clyde Herrick.*

Figure 5-6

Basic plan of an FM automatic frequency control system.

connected across the oscillator coil. The diode has junction capacitance, and the reverse-biased diode "looks like" a small capacitor. If the value of the reverse-bias voltage changes, the diode functions as a variable capacitor. In turn, the oscillator frequency can be controlled (within limits), by variation of the reverse-bias voltage on the varactor diode.

Recall that the output voltage from an FM detector (discriminator or ratio detector), will have a positive or negative polarity, depending on whether the incoming station frequency is above or below the center frequency of the tuned transformer that drives the detector diodes. Suppose that the local os-cillator drifts to a higher frequency than normal. This oscillator drift shifts the normal difference frequency, and the FM detector output voltage becomes more positive than is normal.

As seen in the diagram in Figure 5-6, this abnormally positive DC volt-age from the AFC filter is applied across the varactor diode, with the result that the diode junction capacitance increases. Increased junction capacitance lowers the operating frequency of the local oscillator to virtually its normal value. AFC troubleshooting should be attempted only on a comparative basis (unless the troubleshooter has adequate prior knowledge of the normal circuit action).

BASIC PLAN OF FM/AM INTEGRATED-CIRCUIT RECEIVER

Some FM/AM radio receivers are designed with almost all of the active devices included in a single, large IC chip. Others have the AM devices in an IC chip, supplemented with discrete FM circuitry. Some IC FM/AM receivers provide mono FM operation, and others provide stereo-multiplex FM operation. It is helpful to observe a typical arrangement of external components for a simpler type of FM/AM integrated circuit, as shown in Figure 5-7.

In the figure, an external FM tuner is used and single-channel audio output is provided (unless a supplementary stereo-FM decoder is employed). A 20-pin integrated circuit contains the active devices and some of the incidental resistors. In isolated cases, the manufacturer may provide a socket into which the IC is plugged. If so, this feature simplifies troubleshooting procedures because a preliminary check can then be made by plugging in a known good IC package.

However, most manufacturers solder the IC pins into the PC board. Accordingly, the troubleshooter must ordinarily resort to comparison tests with respect to a similar receiver that is in normal operating condition. Of course, some troubleshooters have prior experience with this type of circuitry, and "know what to look for" in particular models. Beginners can benefit by keeping a notebook with detailed records of receiver models that have been repaired satisfactorily with a listing of quick checks and measured values.

FRONT END WITH SEPARATE OSCILLATOR AND MIXER TRANSISTORS

In the more elaborate types of AM and FM/AM receivers we will find separate oscillator and mixer sections instead of a single autodyne section. A simple example of an AM front end with separate oscillator and mixer circuits is shown in Figure 5-8. It is helpful for the troubleshooter to clearly distinguish between the two basic types of tuner circuitry. Observe in the diagram that transistor Q1 functions as a heterodyne mixer, and that Q2 operates as a local oscillator. The oscillator output is injected at the emitter of Q1, and the incoming station signal is applied to the base of Q1.

Note that winding N3 of T_{osc} has only a small impedance at the station signal frequency, whereby C2 serves as an emitter bypass capacitor for Q1 as well as serving as a coupling capacitor from the local oscillator to Q1. Technically, the advantage of a separate oscillator section is reduced "pulling" of the oscillator frequency by the mixer tuning capacitor, as compared with autodyne

Figure 5-7

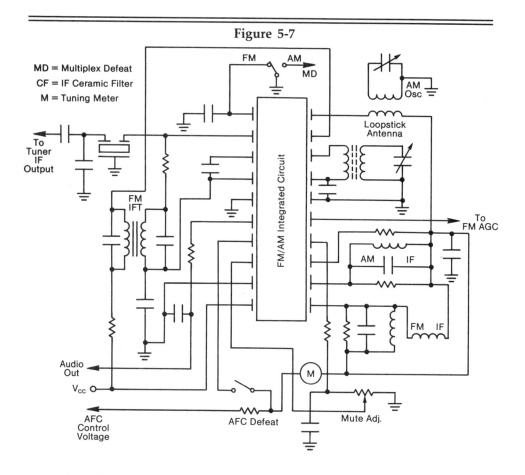

MD = Multiplex Defeat

CF = IF Ceramic Filter

M = Tuning Meter

Note: A ceramic filter is an electrically-coupled two-terminal piezo-electric resonator. Monolithic filters with ceramic substrates are also called ceramic filters.

Note: This is one of the simpler standard arrangements of the external components for an FM/AM integrated circuit. Observe that an external FM tuner is used, and an external stereo-multiplex decoder may be employed, if desired. All of the AM active devices are included in the IC package. An automatic frequency control (AFC) defeat switch is provided, as explained previously. A mute threshold control is also provided to eliminate or minimize noise when tuning from one FM station to another. Note that a ceramic IF filter functions as a tuned IF transformer; it is less costly (and not as selective) as tuned circuitry.

Figure 5-7 (Continued)

Troubleshooting procedures should be limited to comparison tests, unless the troubleshooter is familiar with the circuit actions.

Figure 5-7

A typical arrangement of external components for an FM/AM integrated circuit.

Figure 5-8

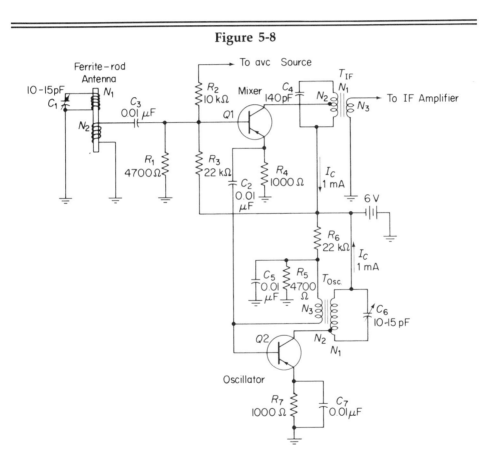

Note: This is a basic example of an AM radio receiver front end with separate mixer and oscillator sections. Its chief advantage is improved oscillator frequency stability under conditions of signal-strength variation. It has a further advantage in that AVC voltage can be applied to the mixer to avoid distortion of strong incoming station signals. Inso-

Figure 5-8 (Continued)

far as troubleshooting procedures are concerned, quick checks are much the same for this arrangement and for the autodyne arrangement.

Technical Note: Coils are often quick-checked for resistance values. An antenna coil has a typical resistance of 1 ohm. A 455-kHz transformer primary has a typical resistance of 6 ohms and the secondary has a resistance of less than 1 ohm. This test is chiefly useful to determine whether a winding may be open or shorted.

Reproduced by special permission of Reston Publishing Company from Radio Theory Servicing *by Clyde Herrick.*

Figure 5-8

An example of an AM front end with separate oscillator and mixer transistors.

circuit operation. "Pulling" is undesirable because it reduces the effective stability of the local oscillator.

Observe in Figure 5-8 that AVC (AGC) bias voltage is applied to Q1 via R2; this bias voltage is signal-developed and reduces the conversion gain as the signal strength increases. The advantage of mixer AVC is prevention of front-end overloading on strong incoming signals. (Overloading results in distortion.) Note that AVC cannot be used in an autodyne circuit inasmuch as a bias-voltage change would cause the oscillator to change its frequency due to inherent varactor action of the converter transistor.

IS THE FM CONVERTER WORKING?

During preliminary troubleshooting procedures, the question is routinely asked "Is the FM converter working?" If the converter is working, but there is no sound output from the receiver, the troubleshooter knows that the fault will be found in the IF section (or following circuitry), and that the FM front end is cleared from suspicion. This quick check can be made quickly and easily, as shown in Figure 5-9. A short-wave AM radio receiver is used as an indicator.

Observe that when an AM radio receiver is tuned to 10.7 MHz, and the end of the whip antenna is placed near the converter coil, or the first FM IF transformer, station signals will be heard from the AM receiver if the FM converter is working. That is, when the FM receiver is tuned to an FM station, and the end of the antenna from the AM receiver is placed near the converter coil, the FM station will be reproduced by the AM receiver if the FM converter section is working.

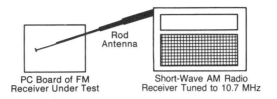

PC Board of FM
Receiver Under Test

Short-Wave AM Radio
Receiver Tuned to 10.7 MHz

Note: This quick check for FM converter operation is based on stray capacitance coupling from the FM converter coil to the end of the rod antenna from the short-wave radio receiver. The antenna tip is placed near the FM converter, and the FM receiver is tuned to any FM station signal. In turn, if the FM converter section is working, the FM station will then be heard from the speaker of the AM radio receiver. On the other hand, if the FM converter is not working, the AM speaker will remain silent. Observe that the AM receiver is tuned to the intermediate frequency of the FM receiver. If the converter section in the FM receiver is working, it will generate a 10.7-MHz signal output corresponding to the incoming FM station signal. This FM signal output is partially changed into an AM signal as shown in Figure 5-10.

Figure 5-9

**Quick check for FM converter operation is made
with a short-wave AM radio receiver.**

The principle of this signal-tracing test is shown in Figure 5-10. Observe that when the FM receiver is tuned slightly off-station, the FM signal carrier then falls on the side slope of the front-end frequency-response curve. In turn, slope detection occurs (the converter output increases as the FM modulation swings up in frequency, but the converter output decreases as the FM modulation swings down in frequency). Consequently, if there is 10.7-MHz output from the FM converter, it now contains both FM and AM modulation. The AM modulation is picked up by the short-wave AM receiver, and the FM station will be heard from the speaker in the AM receiver.

In case a 10.7-MHz signal is found at the FM converter output, the next question is whether there is an output signal from the FM IF section. To determine this, place the end of the rod antenna from the short-wave AM radio receiver near the ratio-detector circuit in the FM receiver. If there is a 10.7-MHz input signal from the FM IF section to the ratio detector, the incoming FM station signal will be heard at a considerably greater volume than before. Or, if there is no output, the troubleshooter concludes that the fault will be found in the IF section of the FM receiver.

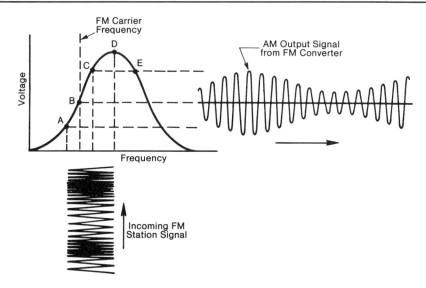

Note: In the slope-detection process, the carrier frequency of the incoming FM station signal is tuned to point B on the FM front-end frequency-response curve. In turn, the FM signal "swings" up and down from B to C to B to A to B. Consequently, the FM converter output now has an AM output signal component that can be "picked up" by a 10.7-MHz AM radio receiver. Observe that the quality of the slope-detected signal is not hi-fi, inasmuch as the side of the front-end frequency-response curve is nonlinear. Also, the slope-detected signal tends to be more or less noisy, since no limiting action is provided by the AM receiver.

The incoming FM station signal has a comparatively high frequency, such as 100 MHz. However, the AM output signal has a low frequency (455 kHz). If the incoming FM signal has a frequency of 90 MHz, the AM output signal will still have a frequency of 455 kHz. These frequency relations are the result of heterodyne action in the FM converter.

Figure 5-10

**Change of frequency modulation (FM) into
amplitude modulation (AM) by slope detection.**

When there is 10.7-MHz input signal to the ratio detector, but no sound output from the FM receiver, the malfunction will be localized either to the ratio detector or to the audio section. To check ratio-detector operation, use a mini-amp/speaker as a signal tracer at the ratio-detector output. If there is sound output from the mini-amp/speaker, it is evident that the trouble will be found in the audio section of the FM receiver. However, if there is no sound output from the mini-amp/speaker, a defect will be found in the ratio-detector circuit.

PC BOARD LAYOUT PRINCIPLES

When troubleshooting AM and FM receivers without service data, pre-liminary "size up" of PC boards can often be facilitated by recognizing the basic principles of layout. In other words, as exemplified in Figure 5-11, an AM receiver board has the detector diode placed at a distance from the loopstick antenna and front-end components. Similarly, as seen in Figure 5-12(a), an FM receiver board has the FM detector section placed at a distance from the front-end components.

In the case of a small FM/AM radio receiver [see Figure 5-12(b)], both the FM and AM front-end components are usually grouped together at one end of the PC board, and both the FM and AM detector components are generally located in an area near the other end of the PC board. Proximity of FM and AM front-end components, and proximity of FM and AM detector components is practical inasmuch as the FM components are "dead" while the receiver is on its AM function, and vice versa. (See also Figure 5-13.)

Figure 5-11

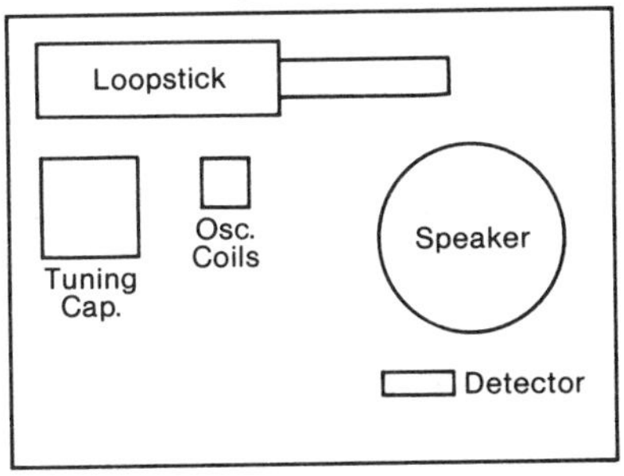

Note: This typical layout for a small AM radio receiver PC board shows the front-end components and their spatial relation to the detector. The layout illustrates the general principle that the low-level high-frequency signal section is placed at a distance from the high-level lower-frequency signal section. (The loopstick antenna is placed comparatively far from the detector.) This placement functions to

Figure 5-11 (Continued)

minimize feedback of detector harmonics to the loopstick, which would produce "birdies" and unstable operation. This basic principle of layout often serves as a helpful guide in preliminary "size up" of a PC board.

Technical Note: A PC board for a small radio receiver is often secured to the case by ferrous self-tapping screws. If a screw mysteriously "disappears" during disassembly, look on the back of the speaker—the screw may have been whisked away to the permanent magnet of the speaker.

Figure 5-11

General plan of PC board layout for a small AM radio receiver.

Figure 5-12

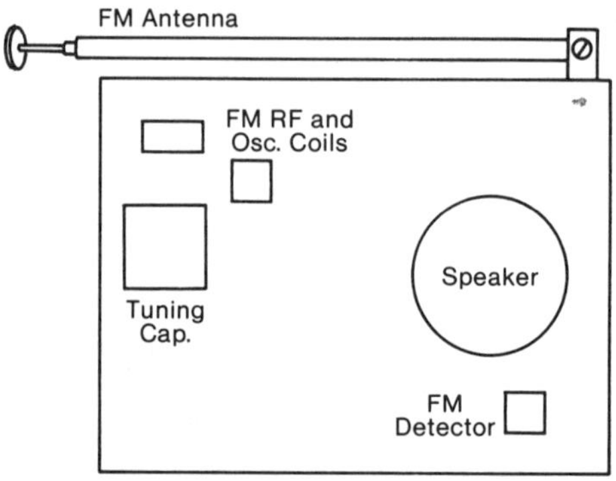

Note: This typical layout for a small FM radio receiver PC board shows the front-end components and their spatial relation to the FM detector. Observe that the layout plan is similar to the previous illustration for an AM receiver. In other words, the low-level high-frequency signal section is placed at a distance from the high-level lower-frequency signal section. The FM detector and preceding FM IF stage generate harmonics, which if allowed to gain access to the front end, would interfere with various incoming FM station signals. From the troubleshooting viewpoint, this basic principle of layout often serves as a helpful guide in preliminary "sizing up" of a PC board.

Figure 5-12 (Continued)

An FM/AM receiver includes the front-end components for both AM reception and for FM reception. An AM/FM switch is provided which applies V_{cc} voltage to either the AM front-end transistors, or to the FM front-end transistors. This is a helpful point to keep in mind when "sizing up" a PC board. For example, if the switch is thrown to its AM position, then a voltmeter will quickly show which front-end transistors serve the AM function, and which transistors serve the FM function.

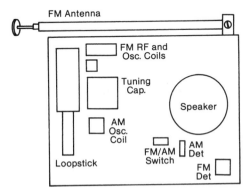

Note: This typical layout for an FM/AM radio receiver PC board shows how the front-end and detector components are arranged. Observe that an FM/AM switch is included in the circuitry. If a milli-ammeter is connected in series with the V_{cc} supply, it will be observed that more current is normally drawn on the FM function than on the AM function, because at least one more transistor is used in FM operation. For example, the current demand on the AM function may be 7.5 mA, whereas the current demand on the FM function may be 9.5 mA. Observe also that the current demand increases considerably as the volume control is advanced. This increased current is drawn by the class-B or class-AB audio output stage. Thus, the current demand typically doubles as the volume control is advanced from low-level to moderate-level audio output. Troubleshooters can sometimes obtain helpful clues concerning fault location by checking the V_{cc} current demand under various conditions of operation.

Figure 5-12

Typical PC board layouts: (a) input and output signal components for a small FM receiver; (b) input and output components for an FM/AM receiver.

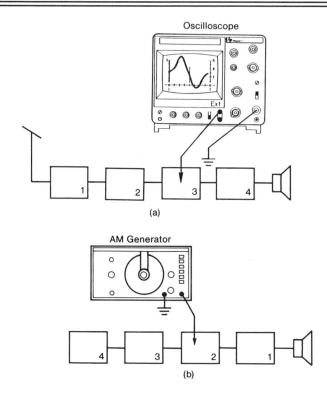

Courtesy, B&K PRECISION, Div. of Dynascan Corp.

Figure 5-13

Signal tracing and signal substitution: (a) signal tracing with the oscilloscope; (b) signal substitution with an AM signal generator.

"DETECTOR SPOTTER" OPERATION OF AM RADIO RECEIVER

When troubleshooting AM radio receivers without service data, it is sometimes desirable to have a "detector spotter" quick check available. For example, if there is no sound output from the receiver, and it has been determined that the autodyne circuit is operating, the next "landmark" of interest is the detector. If the detector diode happens to be "buried," or if there are several visible diodes, then a "detector spotter" quick check is very handy.

This quick check is easily made, as shown in Figure 5-14. It requires only a miniclip lead. One end of the lead is draped over or around the loopstick

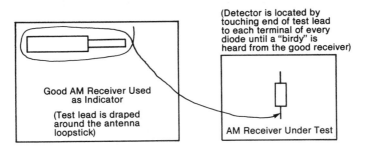

Note: When the test lead is draped around the loopstick, "gimmick" coupling is provided to the antenna section of the good receiver. The second harmonic of the IF signal (910 kHz) will be coupled via the test lead. Similarly, the third harmonic of the IF signal (1365 kHz) will be coupled via the test lead. Harmonics represent spurious signal inputs to the good receiver, with the result that "birdies" or heterodyne squeals are produced. Observe that only the detector diode in the receiver under test can cause "birdies" inasmuch as this is the only diode that processes the 455-kHz IF signal.

In most situations, diode terminals are accessible from the component side of the PC board with a miniclip. However, when a diode cannot be accessed from the component side, it is always possible to access the diode terminal from the solder side of the board.

Figure 5-14

Arrangement of the "detector spotter" quick checker.

antenna in the good receiver, and the other end of the lead is used as a probe. The tuning dial of the good receiver is set to approximately 910 kHz (second harmonic of the IF frequency). Then, the "probe" is touched successively to each terminal of each diode on the PC board in the receiver under test. When the input terminal of the detector diode is touched, a heterodyne squeal ("birdy") will be heard from the speaker. Other "birdies" will be heard at other points on the tuning dial. Observe that this quick check is applicable only when the detector diode in the receiver under test is OK. (If the diode is open or short-circuited, no "birdies" will be heard.)

Observe that the basis of this quick check is harmonic generation by the detector diode. In other words, when the IF signal is applied to the detector diode, half-wave rectification occurs, and the secondary of the last IF transformer accordingly "sees" a very high impedance on one half cycle, and a very low impedance on the other half cycle. The result is substantial distortion of the IF waveform, accompanied by generation of a second harmonic, third harmonic, and so on. The second harmonic of the IF frequency ($2 \times 455 = 910$ kHz) is the strongest harmonic, and it produces the loudest "birdy" in the good receiver.

CONVERTER VOLTAGES IN OSCILLATORY AND NONOSCILLATORY STATES

A practical example of DC voltage relations in converter circuitry is shown in Figure 5-15. The oscillator tuning capacitor was short-circuited in this example, and the receiver was dead, although a hissing sound could be heard from the speaker when the volume control was turned up to maximum.

Comparative DC voltage measurements at the converter transistor terminals in the bad receiver and the good receiver were as follows:

Bad Receiver	Good Receiver
Collector, 4.81 V	Collector, 4.80 V
Base, 2.11 V	Base, 2.09 V
Emitter, 1.45 V	Emitter, 1.55 V

The emitter voltages show the greatest discrepancy, although the difference is not alarming.

However, the comparative voltages provide a definite trouble clue when bias values are noted. Thus, the base-emitter converter bias in the bad receiver is 0.66 volt, whereas the base-emitter converter bias in the good receiver is 0.54 volt. These bias voltages are evaluated as follows:

1. The bias voltage on the converter transistor in the bad receiver is in the conduction region.

2. The bias voltage on the converter transistor in the good receiver is in the cutoff region.

3. The converter transistor in the good receiver only *appears* to be cut off; it is not actually cut off because the local oscillator is operating, and the transistor conducts on the positive peaks of the oscillator waveform.

4. The converter transistor in the bad receiver is *not* cut off; *it is biased for class-A amplification.* Because the transistor is evidently workable, the logical conclusion is that there is a fault in the local-oscillator circuit, which kills oscillation.

5. A follow-up RFI test between the two receivers confirms that the local oscillator in the bad receiver is either dead or far off frequency.

6. As noted above, the fault in the local-oscillator circuit was a short-circuited tuning capacitor.

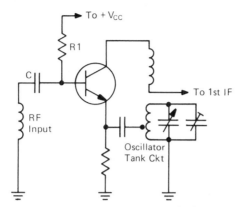

TROUBLE SYMPTOM: "DEAD RECEIVER"

Note: The base of the converter transistor is forward-biased by R1. However, this forward bias is masked by charge build-up on C when the local oscillator is operating. Oscillation normally occurs because the collector is inductively coupled to the oscillator tank, and positive feedback sustains oscillation.

This oscillation is normally sufficiently strong that the transistor conducts heavily on peaks of the oscillator waveform. Peak conduction results in spurts of electron flow in the base circuit; some of these electrons are stored on the right-hand side of C, and the average bias on the base becomes less positive.

In this particular example, oscillation caused the average bias on the base to be 0.54 volt (in the cutoff region).

When the oscillator tuning capacitor became short-circuited, oscillation was killed, and the bias on the base assumed a value of 0.66 volt (in the conduction region).

Figure 5-15

Typical oscillator circuitry in a converter stage.

ALL VOLTAGES NEGATIVE

In the example above, all of the voltages were positive, inasmuch as the negative side of the power supply was connected to ground (common). However, in the following example, all of the voltages are negative with respect to ground (common). The basic DC-voltage distributions that are involved in these examples are depicted in Figure 5-16.

In the example following, the bad receiver had the same fault as in the previous example—the local oscillator was killed, (due to a solder bridge that short-circuited the oscillator coil). Comparative DC voltage measurements at the converter transistor terminals in the bad receiver and in the good receiver were as follows:

Bad Receiver	Good Receiver
Collector, −7.96 V	Collector, −7.93 V
Base, −2.31 V	Base, −2.24 V
Emitter, −1.67 V	Emitter, −1.77 V

Note that the receiver in this example employed PNP transistors; $+V_{cc}$ was returned to ground (common), and all of the transistor terminal voltages were negative with respect to ground.

The differences in DC voltages at the converter transistor terminals in the bad receiver and good receiver are not alarming. As in the case of the first receiver, however, the comparative voltages provide a definite trouble clue when bias values are noted. Observe that the base-emitter bias was 0.47 volt in the good receiver, whereas the base-emitter bias was 0.64 volt in the bad receiver.

The trouble clue is evident in the fact that the converter transistor was biased into the cutoff region in the good receiver, whereas the converter transistor was biased into the conduction region in the bad receiver.

To recap the trouble analysis in this example, note that:

1. The converter transistor in the good receiver only *appears* to be cut off. It is cut off *on the average*, as "seen" by a DC voltmeter. In fact, the transistor conducts in spurts on the peaks of the oscillator waveform.

2. The pulse-type conduction of the converter transistor in the good receiver stores a charge on the base coupling capacitor, and on the emitter coupling capacitor. These stored charges oppose the fixed bias and shift the *average apparent bias* into the cutoff region.

3. There is no pulse-type conduction of the converter transistor in the bad receiver, and only the fixed bias voltage is measured. This fixed bias is 0.64 volt. In other words, the converter transistor in the bad receiver is biased for class-A amplification.[1]

---

[1]Converter circuit action is obtained only when the converter transistor operates in class AB, class B, or class C; no beat output can occur in class A.

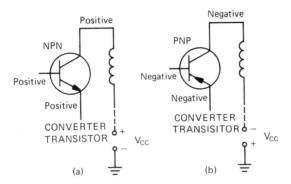

(a)   (b)

Note: All of the NPN transistor terminals are positive, because the negative side of the power supply is connected to ground. On the other hand, all the PNP transistor terminals are negative, because the positive side of the power supply is connected to ground.

Although this is the common practice, it is not the invariable rule, and the troubleshooter should keep the following exceptions in mind:

1. Occasionally, an NPN transistor will be operated with the positive side of the power supply returned to ground. In such a case, all of the NPN transistor terminals will be negative with respect to ground.
2. Occasionally, a PNP transistor will be operated with the negative side of the power supply returned to ground. In such a case, all of the PNP transistor terminals will be positive with respect to ground.

A converter transistor operates nonlinearly (a beat-frequency output cannot be produced if the transistor operates linearly). Therefore, the measured terminal voltages are affected, not only by the oscillator waveform, but also by the RF input signal waveform. This is just another way of saying that the bad receiver should be tuned to the same frequency as the good receiver.

Figure 5-16

Basic DC voltage distribution: (a) all voltages positive with respect to ground; (b) all voltages negative with respect to ground.

4. This circuit-action analysis leads the troubleshooter to the conclusion that the converter transistor is probably in working condition, and that some circuit fault is killing the local-oscillator action.

OSCILLATOR SIGNAL SUBSTITUTION

Oscillator signal substitution is commonly provided by connecting the RF output from a signal generator to a small coil, and coupling this coil to the antenna loopstick in the bad receiver.

In turn, when the receiver is tuned to an AM station, and the generator is set to the station frequency plus 455 kHz, reception will be obtained in the event that the converter transistor is workable, but the local-oscillator function is dead. Preliminary conclusions can therefore be confirmed or rejected.

Similarly, the receiver may be tuned to an FM station, and the output from an RF generator coupled into the FM mixer circuit. Note that the difference frequency is 10.7 MHz in this situation. This oscillator signal-substitution test should be made only with a lab-type VHF signal generator, however.

SECOND-HARMONIC GENERATION BY FM DETECTOR

An instructive FM circuit-action demonstration can easily be made with an AM short-wave radio receiver. A clip lead is connected at one end to the rod antenna on the AM receiver, and the other end of the clip lead is capacitively coupled to a diode in the FM detector. (The clip may be brought near a diode lead, or it may be clamped to the insulated body of the diode.) An FM station is tuned in on the FM receiver, and the volume control turned down. Then, when the AM receiver is tuned to 21.4 MHz, and its volume control is advanced, the FM station will be heard from the AM receiver. (The FM tuning control may need slight adjustment to obtain optimum slope detection.)

AUTOMOBILE RADIOS

Troubleshooting automobile radios without service data is basically the same as for portable and table-model AM and FM receivers, except for "packaging" considerations. The environment of a car radio is such that the antenna is a ready suspect in case of weak, noisy, or no-signal trouble symptoms. Check the following possible faults:

1. Poor or broken connection of antenna cable to antenna or to plug on radio case
2. Faulty ground connection at base of antenna
3. Damaged and/or wet antenna cable

If a new antenna is installed, remember that the alignment of the front end may have been disturbed. To ensure optimum reception, adjust the antenna trimmer for maximum output with the tuning dial set to a weak station in the 1400 kHz region. This antenna trimmer capacitor will be located in the vicinity of the tuning shaft, often behind the tuning knob, or possibly near the antenna connector.

In the case of a "dead radio" trouble symptom, the fuse is a ready suspect. This fuse will be located near the radio, or on the firewall, and will be clipped into a fuseholder. The current rating of the fuse will be marked on the end of the fuse and/or holder. Usually, replacement of a blown fuse will restore the radio to normal operation. However, if the replacement fuse blows, the troubleshooter can conclude that there is a fault in the radio circuitry that is causing excessive current demand.

The environment of a car radio also results in more speaker malfunctions than home radios do. For example, the troubleshooter may find warped speaker cones, damp voice coils, or foreign matter in the voice-coil gap. In the case of small speakers that have been operated at excessively high volume levels, look for evidence of overheating in the voice coil and its leads. If only one speaker is provided, it can be temporarily disconnected and a substitution test can be made with a known good speaker. In a stereo system, cross-checks will quickly show whether the trouble is in a speaker or inside the radio.

When the fault is inside the radio, the troubleshooting procedure is the same as that for AM and FM receivers discussed in Chapters 4, 5, and 6. The only exception is that the power-supply voltage must be provided by an automobile battery or by an equivalent bench power supply.

IC FUNCTIONAL OVERVIEW

A widely used IC arrangement of an RC phase-locked loop (PLL) stereo decoder is shown in Figure 5-17. It differs from the IC described previously in that it does not require any tuned circuit. This IC employs a PLL and dividers instead of the 76-kHz resonant circuit noted in the previous chapter. Also, the arrangement in Figure 5-17 does not require synchronous detectors. The external components shown in Figure 5-17(a) consist of an RC network for the VCO (voltage-controlled oscillator) which is the heart of the PLL, and also another RC circuit for the loop filter, and a single capacitor for the switch filter.

This IC also provides an output of the 19-kHz pilot signal, an indicator, and the L and R outputs. Its performance characteristics are, for the most part, superior to the stereo decoder described previously. Basic functional parameters are:

1) *Input impedance.* Typically 50 kilohms.
2) *Channel separation.* Denotes the relative interference reduction between the two audio output channels (40 dB is typical).

Figure 5-17

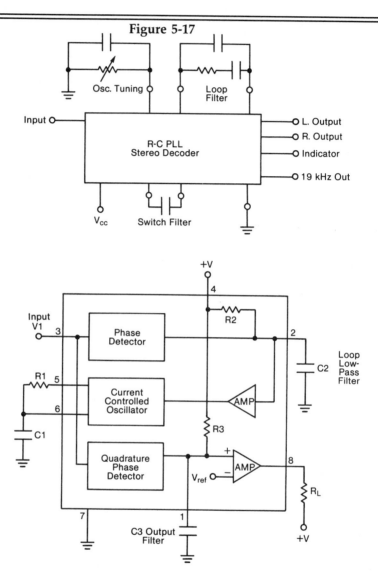

Note: When the phase difference between the VCO and the input signal is constant, the phase loop is locked. If either the input or reference signal, or the VCO output, changes in phase, the phase detector and filter produce a DC error signal that is proportional in magnitude and polarity to the original phase change. This error voltage changes the VCO phase, so that it again locks into the reference signal. Since the error voltage is brought out, as well as the VCO

Figure 5-17 (Continued)

frequency controlling network, the VCO signal and the reference input, a number of different functions can be performed by the PLL IC.

Figure 5-17

Arrangement of an RC phase-locked loop (PLL) stereo decoder: (a) IC with external components; (b) inside the IC.

3) *Audio output voltage.* Denotes the output signal value at either stereo channel for full limiting with minimum input signal. (A typical rating is 500 rms mV.)

4) *Channel balance.* Denotes the relative amplitude difference between the two stereo channels. A typical value is 0.1 dB with a maximum limit of 1.5 dB.

5) *Capture range.* Rated deviation from the center frequency over which stereo demodulation will take place. (A typical rating is ±3.5 percent.)

6) *Total harmonic distortion.* Rated sum of all distortions at all harmonic frequencies. (A typical rating is 0.3 percent.)

7) *19 kHz rejection.* Expresses the relative attenuation of this frequency in the audio output channel. (A rating of 34.5 dB is typical.)

8) *38 kHz rejection.* Expresses the relative attenuation of this frequency in the audio output channel. (A rating of 45 dB is typical.)

Next, an IC arrangement for a clock circuit is shown in Figure 5-18. Some radio receivers and many VCRs contain clocks. This IC provides all of the functions required in an electronic clock, whether operated from a 60-Hz power source, from an automobile battery, boat battery, or plane battery. In various applications, it can accept inputs from a color-TV crystal oscillator, or from an AC power line. These signals are then counted down in the IC to generate individual minutes, ten minutes, and hour displays.

A 3.75-Hz signal is available from the IC for flashing specific numerals or messages. Separate leads are brought out from the segment drivers to each of the LEDs or fluorescent numeral indicators. Only three control inputs are required. To change any particular digit, the "increment" input permits the operator to select hours, tens of minutes, minutes, or, to let the clock run. Once the desired state is selected, the digit can then be incremented by providing one impulse or pushbutton closing for each advance in the selected digit. The "re-

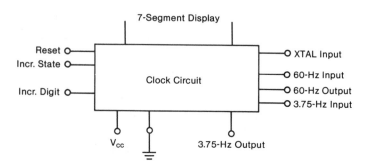

7-Segment Display

Reset
Incr. State

Clock Circuit

Incr. Digit

XTAL Input
60-Hz Input
60-Hz Output
3.75-Hz Input

V_{cc}

3.75-Hz Output

Reproduced by permission of Prentice Hall From Complete Handbook of Practical Electronic Reference Data *by Walter H. Buchsbaum.*

Figure 5-18

An IC arrangement for a clock circuit.

set" input automatically resets the clock to indicate 1:00. Basic functional parameters are:

1) *Power supply voltage.* A nominal voltage of + 5V is typical.
2) *Logic control levels.* For logic 1, a typical range is 2 to 5 V. For logic 0, a typical range is 0 to 0.3 V.
3) *Maximum power dissipation.* With all segments illuminated, approximately 500 mW is required.

6

PROGRESSIVE RADIO TROUBLESHOOTING TECHNIQUES

Signal-Injection Procedures • AM/VHF Aircraft Pocket Radio • Modulation vs. Demodulation • Audio Section Malfunction • The "Sound of Overmodulation" • Measurement of Quench Frequency • Measurement of High-Frequency Voltage at Antenna • Scanner-Monitor Radio Receivers • Incidental Frequency Modulation • Quick Check for AM Rejection by FM Receiver • CB Walkie-Talkie RFI • Check of Modulation Waveform • Test Tip Reminders • Concluding Points • IC Functional Overview

SIGNAL-INJECTION PROCEDURES

It is frequently helpful, or necessary, to inject a signal voltage at a selected point in a circuit when troubleshooting radio receivers without service data. Signal injection procedures involve selection of a circuit point that will give a clear-cut answer to the troubleshooter's question. Also, consideration is sometimes required concerning suitable injection techniques. In other words, attention should be given to circuit disturbance hazards and their avoidance. Beginners need to keep in mind that a signal-injection procedure that is satisfactory at audio frequencies may be poor practice at radio frequencies. Moreover, a procedure that is satisfactory at AM broadcast radio frequencies may be far off-target at FM broadcast radio frequencies.

Observe that a blocking capacitor was not used in series with the output lead from the audio oscillator in Figure 6-1. A blocking capacitor is not needed

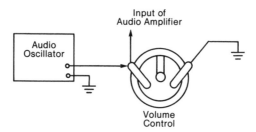

Note: The volume control is prominently located at the edge of the PC board in an AM or FM radio. It is provided with two long solder lugs, one of which is connected to ground; the other lug is connected to the input of the audio amplifier. To inject an audio signal, the output leads from an audio oscillator are clipped between the "hot" volume-control lug and ground. (Ground will be one terminal of the V_{cc} supply; usually $+V_{cc}$ if PNP transistors are used, or $-V_{cc}$ if NPN transistors are used.)

Figure 6-1

The volume control is a prominent "landmark" for audio signal injection.

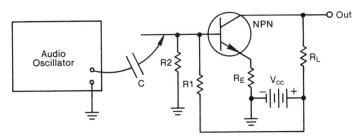

Note: In this example, the signal voltage is being injected at a circuit point which operates above DC ground potential. (R1 and R2 form a base bias-voltage divider.) In turn, a blocking capacitor C is required in series with the "hot" output lead from the audio oscillator to avoid drain-off of base bias voltage via the test lead and the attenuator in the audio oscillator. If the blocking capacitor is omitted, the stage is forced to operate in class B due to low output resistance of the audio oscillator. Accordingly, the output signal from the stage would be greatly distorted, and the test conclusions would be false.

Signal injectors called pencil noise generators have been used extensively in the past for both audio-amplifier and AM radio testing. However, they have lost favor, and are seldom used by technicians. Instead, audio oscillators are preferred for signal injection into audio circuitry, due to the control of signal level that is provided and also to the control of test frequency that is available. Observe that although AM signal generators have an audio test signal output that is adjustable in level, the signal frequency is fixed (typically 400 Hz).

Figure 6-2

Blocking capacitor is required when signal-injection point operates above DC ground potential.

here, inasmuch as the volume control is returned directly to ground. However, consider next the injection of an audio signal at the output stage of the amplifier, as shown in Figure 6-2. Now, the audio signal is being injected at a circuit point which is above DC ground potential. Therefore, a series blocking capacitor must be included to avoid drain-off of DC bias voltage back through the audio-oscillator attenuator.

If an RF signal is to be injected into the loopstick antenna of an AM broadcast receiver, direct connection to the circuitry is not used. Instead, inductive coupling is provided from the AM signal generator to the loopstick, as seen in Figure 6-3. (This method of signal injection imposes minimum disturbance of the antenna input circuit—an essential consideration when alignment with test equipment is planned.) Inductive coupling is provided by means of a "radiation loop," which consists of a few turns of wire several inches in diamater, and placed a foot or more from the receiver.

Next, if an RF signal is to be injected at the rod antenna of an FM broadcast receiver, direct connection to the circuitry is not good practice. Instead, capacitive coupling ("near-field radiation") is provided from the FM signal generator to the rod antenna, as depicted in Figure 6-4. Signal coupling is obtained by means of a "radiation wire" several inches long and placed at a moderate distance from the rod antenna. This method is equally useful for stereo, mono, or sweep-alignment tests.

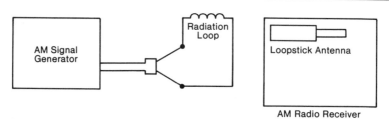

Note: To inject a test signal from an AM generator into the front end of an AM broadcast receiver, the approved procedure is to employ a radiation loop consisting of a few turns of wire, and placed a foot or two from the receiver. Circuit disturbance and detuning effects are therefore minimized. A standard test signal uses 30 percent 1-kHz modulation of the carrier. Observe that if the test signal is to be used for precise alignment with instruments, a DVM is also connected across the speaker terminals and the RF signal level is reduced as much as possible while still obtaining adequate output reading on the DVM. (This ensures that signal circuits in the receiver will not be overloaded, and that the circuits will be aligned for optimum weak-signal sensitivity.)

An AM signal generator provides RF CW or amplitude-modulated output over a wide range of frequencies. Its calibration can be checked against WWV signals, or against broadcast-station frequencies (if known).

Figure 6-3

Inductive coupling is used for signal injection to the loopstick antenna of an AM broadcast receiver.

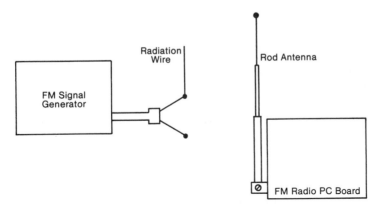

Note: To inject a test signal from an FM generator into the front end of an FM broadcast receiver, several inches of "radiation wire" may be connected to the "hot" output lead from the FM signal generator and placed a foot or two from the rod antenna of the FM receiver. The test signal may be either a CW (single-frequency continuous-wave voltage), or a frequency-modulated carrier, such as a 100-MHz carrier frequency modulated by a 1-kKz audio signal. Again, for sweep-alignment purposes, the test signal may have a center frequency of 100 MHz, for example, which is swept at a 60-Hz repetition rate back and forth through the complete FM channel. (When sweep alignment is used, an oscilloscope must be connected at the FM detector output instead of a DVM.)

Although the incidental frequency modulation present in economy-type AM generator output signals can be used for go/no-go signal-injection tests, alignment procedures require the use of an accurate FM signal generator or FM sweep-marker generator.

Figure 6-4

Capacitive coupling is used for signal injection to the rod antenna of an FM broadcast receiver.

AM/VHF AIRCRAFT POCKET RADIO

Note that an AM/VHF aircraft receiver is very similar to an FM/AM broadcast receiver. The primary difference is that the FM tuning range is typically from 108 to 135 MHz, instead of 88 to 108 MHz.

MODULATION VERSUS DEMODULATION

When troubleshooting transceiver equipment without service data, it is often helpful to clearly recognize modulation and demodulation processes. (Demodulation is also called detection.) We will see that the modulation process used in a radio transmitter is closely related to the demodulation process used in a radio receiver. Modulation consists of varying the amplitude of a high-frequency continious (carrier) wave in step with the changing value of an audio-frequency modulating signal. Modulation is required at the transmitter because it is impossible to radiate the audio-frequency modulating signal directly with any practical degree of efficiency. However, the modulated carrier wave can be radiated with good efficiency.

Demodulation is the process whereby the audio-frequency modulating signal is recovered from a previously modulated carrier wave. With reference to Figure 6-5, an audio-frequency modulating signal is shown at (a). Next, a CW radio-frequency carrier is seen at (b). If these two waveforms are simply mixed, modulation does *not* occur. That is, a simple mixture of the two waveforms produces the result depicted in part (c) of the figure. Although the simple mixture of waveforms can be radiated, the original audio-frequency signal will be lost. This is just another way of saying that when the waveform at (c) is radiated from a transmitting antenna, the receiving antenna "sees" the carrier only, as shown in part (b).

A simple mixture of waveforms is produced by a linear mixer, such as a class-A stage. Observe that if the audio waveform at (a) is mixed with the carrier at (b) through a *nonlinear* stage, such as a class-B stage, then a modulated wave is outputted, as shown at (d). In other words, the nonlinear stage causes the amplitude of the radio-frequency carrier to vary in step with the changing value of the audio-frequency modulating signal. It is evident that the receiving antenna will "see" the modulated wave as a carrier which is varying in amplitude.

Next, observe that when the modulated waveform at (d) is passed through a nonlinear stage, such as a class-B stage, one polarity of the waveform will be passed, and the other polarity of the waveform will be rejected. This is the essential basis of demodulation. It is evident that when the output from the demodulator, such as a class-B stage, is processed by a low-pass filter, the original audio-frequency signal will be recovered.

To repeat an essential point, modulation and demodulation are both nonlinear processes. From the viewpoint of the troubleshooter, one important conclusion is that the bias voltages in modulation and demodulation circuitry are a basic concern. A modulating system or a demodulating system must never operate as a linear system. Of course, distortion can (and often does) occur in malfunctioning nonlinear systems. For example, an overdriven device can clip the peaks of a modulated waveform to such an extent that the receiver operator objects to the quality of sound reproduction.

This is another example of the benefit of knowing "the sound of clip-

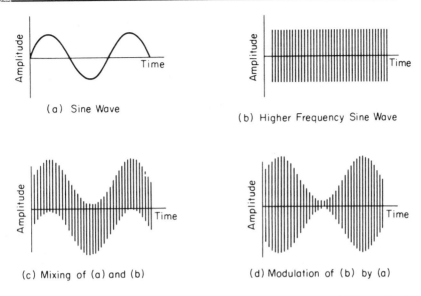

(a) Sine Wave

(b) Higher Frequency Sine Wave

(c) Mixing of (a) and (b)

(d) Modulation of (b) by (a)

Note: Simple (linear) mixing of the waveforms in (a) and (b) results in the waveform shown in (c). Although the higher-frequency component of the waveform can be radiated from the transmitting antenna, the lower frequency (audio frequency) component of the waveform cannot be radiated. In turn, the receiving antenna "sees" only the higher-frequency component, as shown in (b). On the other hand, nonlinear mixing of the waveforms in (a) and (b) produces the modulated waveform shown in (d). This modulated waveform will be radiated completely by the transmitting antenna, and the receiving antenna will "see" the complete waveform.

Reproduced by special permission of Reston Publishing Company from Radio Theory and Servicing *by Clyde Herrick.*

Figure 6-5

Basic waveform in modulator, demodulator, and simple mixer circuits.

ping,"as noted previously. The degree of modulation can be established at the transmitter from zero to 100 percent, and the quality of sound reproduction will be normal up to 100 percent modulation. (The waveform in Figure 6-5(d) is almost 100 percent modulated.) In a malfunctioning modulator, the percentage of modulation might be 150 percent of the peaks of the modulating audio-frequency signal. Overmodulation can result in intolerable distortion because the resultant waveform has "gaps" corresponding to the duration of each over-modulation interval.

It is desirable to have 100 percent modulation when feasible, because the transceiver system then operates at maximum efficiency. On the other hand, overmodulation results in objectionable sound quality, although the system is operating at comparatively high efficiency. Therefore, troubleshooters can benefit by familiarizing themselves with "the sound of overmodulation." In turn, when this malfunction occurs, the troubleshooter recognizes that the modulator section needs to be checked out.

SCANNER-MONITOR RADIO RECEIVERS

Various types of scanner-monitor receivers are in extensive use for high-frequency communication. This type of receiver automatically tunes the front end progressively through the given frequency spectrum. If there is no active station during the scan, the receiver then repeats the scanner action. However, if an active station signal is encountered, the circuitry locks on the signal frequency and holds as long as the signal is present. When the signal ceases, the receiver resumes scanner action.

Typical frequency ranges are low-VHF, from 30 to 50 MHz; ham, from 50 to 54 MHz; government, from 138 to 144 MHz; ham, from 144 to 148 MHz, high-VHF, from 148 to 174 MHz; ham/government, from 380 to 450 MHz; low-UHF, from 450 to 470 MHz; and high-UHF, from 470 to 512 MHz. For example, a simple scanner-monitor covers four channels within the frequency range from 148 to 174 MHz for reception of police, fire, CD, and weather communications and reports.

In the example above, *the scanning speed is 10 channels per second*, with the option of manual channel selection by means of slide switches. After the front end is locked on an active channel, a two-second delay is provided after the signal ceases before scanner action is resumed. This delay serves to prevent missed callbacks. Note that whenever the receiver is tuned to a particular channel, a light-emitting diode (LED) glows and indicates the channel number. Lock-out switching positions enable a "skipper" circuit wherein the scanner bypasses the locked-out channels.

Dual-conversion FM superheterodyne circuitry is utilized in this example. The incoming signal frequency is heterodyned down to an IF frequency, amplified, and then heterodyned down again to a lower IF frequency. This process provides optimum rejection of potential interference. Automatic scanner action is controlled by digital electronic switching facilities. Scanner activity may be described briefly as follows. Figure 6-6 shows a simplified two-channel scanner arrangement. *The scanning rate is determined by a multivibrator (also called the "clock")*. In turn, a bistable multivibrator (also called a "flip-flop") alternately grounds out the channel-9 crystal oscillator, or the 23-channel synthesizer crystal oscillator.

The receiver output is automatically switched back and forth between

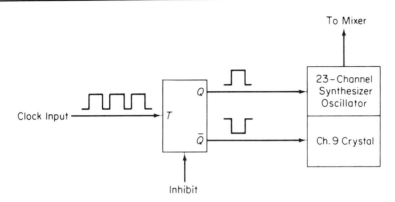

Clock Input

T

Q

$\bar{Q}$

To Mixer

23-Channel
Synthesizer
Oscillator

Ch. 9 Crystal

Inhibit

Note: This simplified partial diagram shows the plan of the scanner section for a 2-channel scanner-monitor radio receiver. From the troubleshooter's viewpoint, it is helpful to observe that a malfunction will usually be localized either to the scanner circuitry or to the front-end circuitry. Insofar as the front end is concerned, its circuitry is basically the same as explained previously for manually tuned receivers. Scanner circuitry is discussed in greater detail subsequently.

Reproduced by special permission of Reston Publishing Company from Radio Theory and Servicing *by Clyde Herrick.*

Figure 6-6

Simplified diagram for a 2-channel scanner-monitor radio receiver.

the two channels. If an incoming signal is present on a channel, the AGC voltage increases and biases off (inhibits) the clock circuit so that the receiver locks on the active channel. Also, the increase in AGC voltage opens the squelch circuit and energizes an LED, as noted previously, to indicate the number of the active channel.

Next, observe the block diagram for a typical VHF-FM scanner receiver as shown in Figure 6-7. Dual-conversion (double-conversion) superheterodyne circuitry is employed. Observe that the local oscillator operates from crystals that are selected by the scanner. *The oscillator frequency heterodynes with the incoming signal frequency to produce an IF frequency of 10.7 MHz. Next, a second oscillator/mixer section heterodynes the first IF signal with an 11.155-MHz frequency to produce the second IF frequency of 455 kHz.* Note that the squelch circuit generates the scanning-control signal (switching voltage). Accordingly, any "open-squelch" signal locks the scanner on the associated channel.

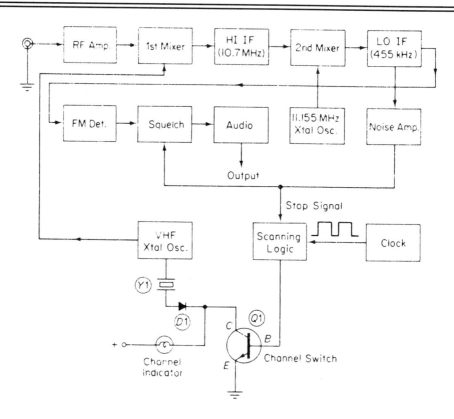

Technical Note: Comparison tests with a known good unit are most helpful when troubleshooting without service data. However, in clocked circuitry, comparison tests are often facilitated by use of the same clock for both the "bad" unit and the "good" unit. This is a technique wherein both clocks are stopped by any suitable means, and a clock-substitution signal is applied to both units. A typical clock-substitution signal is obtained from a square-wave generator that is set to an appropriate output level and repetition rate.

Figure 6-7

Block diagram for a typical VHF-FM receiver.

THE "SOUND OF OVERMODULATION"

It is helpful for the troubleshooter of transceiver equipment to know "the sound of overmodulation" in preliminary evaluation of malfunction. Accordingly, the familiarization demonstration shown in Figure 6-8 can be helpful. The AM signal generated can be externally modulated by the output from an audio oscillator, for example. In turn, as the percentage of modulation is increased over 100 percent, the sound output from the radio receiver becomes typically harsher and loses the timbre of a pure sine wave. It is also instructive to externally modulate the AM signal generator from the output from another AM or FM radio receiver, to observe the change in characteristics of speech and musical passages as the percentage of modulation is increased over 100 percent.

A partial schematic diagram for a scanner circuit is shown in Figure 6-9. Observe that a unijunction transistor (UJT) Q1 functions as a pulse generator and a clock. Sharp pulses are applied to a shaping circuit and to the control section. Consequently, the control circuit interrupts the pulse train when the squelch voltage changes. Uninterrupted (uninhibited) pulse trains are counted by the decade counter. (Counting by digital-logic circuitry is explained in greater detail in a later chapter.)

The arrangement in Figure 6-9 employs a transistor-transistor-logic (TTL) IC chip that counts by tens. Note that *this counting action proceeds in terms of binary numbers—not decimal numbers.* The counter output is given in binary coded decimal numbers (BCD numbers). This BCD signal is applied to a decoder circuit which converts the pulses into suitable form for driving the LEDs. In Figure 6-9, the BCD decoder is a binary-to-octal type that converts BCD numbers into octal numbers. In turn, eight output lines are provided, with one line for each channel of operation.

Observe also another typical arrangement for scanner logic circuitry, as shown in Figure 6-10. This configuration includes two flip-flops and four two-input NAND gates that sequentially select one of four crystals. Note that *a gate is basically an electronic switch.* A NAND gate requires two simultaneous input pulses to switch its state (produce an output pulse), and the output pulse has opposite polarity with respect to the input pulses.

From the troubleshooter's viewpoint, the important voltages in the circuit are the waveforms associated with the scanner activity, as shown in Figure 6-10(b). Observe that the NAND gates are connected to the flip-flops so that they produce a zero output (grounded output) only when both input pulses are present. Note also the waveforms from FF1 and FF2 beneath clock pulse 1. During this interval, the not-Q ($\overline{Q}$) of FF1 and the not-Q ($\overline{Q}$) of FF2 have maximum voltage (logic-high voltage), so they are employed to drive G1.

Q and not-Q (Q and $\overline{Q}$) merely denote opposite pulse polarities. Then, when clock pulse 2 is present, the Q output of FF1 and the not-Q ($\overline{Q}$) output of FF2 are logic-high (all others are logic-low), so they are utilized to drive G2. This sequence of logic-circuit action continues until condition 4 is completed, and then automatically starts over again.

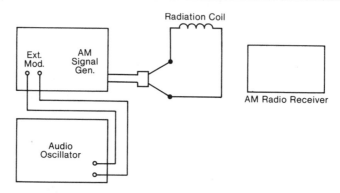

Note: This familiarization demonstration makes use of an audio oscillator to overmodulate an AM signal generator. The output signal is picked up from the radiation coil and is reproduced by an AM radio receiver. The operator can observe the change in timbre that occurs as the AM signal generator is progressively modulated over 100 percent. Alternatively, the audio output from another AM or FM radio receiver may be used instead of the audio oscillator to overmodulate the AM signal generator. In turn, the operator can observe the change in speech and music characteristics that take place in the presence of overmodulation.

Technical Note: Under various trouble conditions, overmodulation will occur in walkie-talkies, citizen's band radios, amateur radio transmitters, and some types of wireless intercoms. Accordingly, it is helpful for the troubleshooter to be familiar with the "sound of overmodulation."

Figure 6-8

**A helpful familiarization demonstration of the
"sound of overmodulation."**

QUICK CHECK FOR AM REJECTION BY FM RECEIVER

An FM IF and detector system normally rejects any amplitude modulation that may be present in an FM signal, as well as any AM station signal. Although it might be supposed that an FM receiver could be checked for AM rejection by means of the signal provided by an AM generator, this is a "tricky"

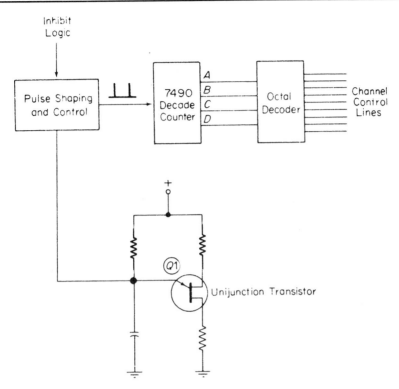

Note: The unijunction transistor operates in a relaxation oscillator circuit, and a sawtooth waveform is generated across the capacitor. This sawtooth wave is then passed through a shaper circuit which outputs narrow pulses of suitable amplitude to drive the decade counter. Observe that octal numbers are base-8 numbers.

Figure 6-9

Partial diagram for a scanner logic circuit.

procedure with many service-type generators, particularly when operated at comparatively high percentages of internal modulation. Incidental frequency modulation can make it appear that an FM receiver has poor AM rejection, although it has normal rejection capability.

However, a reasonably good quick check for AM rejection can be made by using a simple external amplitude modulator, as shown in Figure 6-11. It is apparent that the semiconductor diode functions as a nonlinear device that generates an AM signal from a mixture of an RF carrier and audio-frequency

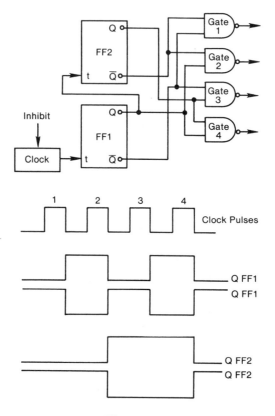

Figure 6-10

**Block diagrams and waveforms for a 4-channel
scanner: (a) arrangement of the digital ICs;
(b) waveforms in the scanning circuits.**

voltages. The advantage of this arrangement is that operation of the oscillator in
the RF generator is practically unaffected by the external modulator activity. The
peak voltage of the AF signal should be about ⅓ the peak voltage of the RF signal
at the diode input.

CB WALKIE-TALKIE RFI

The 27-MHz oscillator is not amplitude-modulated in a typical 300-mW
CB walkie-talkie; amplitude modulation takes place in the final amplifier following the crystal oscillator. Nevertheless, there is normally a small amount of
incidental frequency modulation. There is also some harmonic output. In turn,

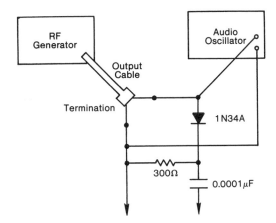

Note: This external modulator arrangement provides practical isolation between the diode resistance variation and the oscillator in the RF generator. It is suitable chiefly for AM rejection checking; the envelope of the modulated waveform is not a true sine wave. Consequently, the arrangement should not be used as a general-purpose amplitude modulator.

Figure 6-11

External modulator arrangement for quick check of AM rejection by FM receiver.

when the normally operating walkie-talkie is located near an FM receiver, transmission is picked up at 108.5 MHz on the tuner (the fourth harmonic of the walkie-talkie Channel 14 transmitting frequency).

Excessive RFI points to a fault in the final-amplifier or modulator circuitry. As noted previously, one of the most common causes of excessive RFI from walkie-talkies is overmodulation and resultant "sideband splatter." Oscilloscope checks in the bad and good walkie-talkies can provide highly informative test data in the case of RFI trouble symptoms.

TEST TIP REMINDERS

Although these test tips are not really new, they are of basic importance in practical troubleshooting procedures, and are recapped as reminders:

Distorted Modulation: Look for a malfunction in the modulator section. The microphone could be defective—try another known good microphone.

Transmitter Operates with XTAL Unplugged: The oscillator, driver, or final circuits are self-oscillatory. The most likely culprit is an open decoupling or bypass capacitor in these sections.

Transmitter Is Off-Frequency: This malfunction is almost certainly caused by a defective quartz crystal.[2]

Weak Transmitter: RF amplifier transistors are ready suspects, followed by defective capacitors. (The final transistor can be damaged if the rod antenna is not fully extended.)

Intermittent Reception: Intermittents may be thermal, mechanical, or triggered by transient voltages. Check switches, defective insulation, solder joints, pressure contacts, and variable controls. Transistors, diodes, resistors, and crystals may become internally intermittent.

Overload on Strong Signals: When the receiver overloads and distorts on strong-signal reception, the AVC circuit is most likely to be defective. Leaky capacitors are the most common culprits.

Weak Reception: When reception is weak, the trouble could be as simple as a poor antenna connection; otherwise, look for a low-gain stage or a defective detector diode. Signal-injection tests made on a good transceiver for comparison with responses of the bad transceiver are invaluable.

Another test tip is that confusion in diagnostic procedures is likely to occur in double-conversion superheterodyne circuitry unless the troubleshooter recognizes the two injection-frequency sections. That is, the RF amplifier is followed by a mixer that uses a variable injection frequency, such as from 2.19 to 31.65 MHz. This first mixer is followed by a second mixer that operates with a fixed injection frequency, such as 1700 kHz. Then, the output from the second mixer is applied to the second detector.

This elaboration of the basic superheterodyne arrangement is employed in various two-way receivers, in CB receivers, and in the more sophisticated types of communications receivers. In the case of a CB receiver, the first injection frequency is adjustable in steps (not continuously variable). The injection frequency is changed in discrete steps by means of a switch control. If a heterodyne *frequency synthesizer* is used, the adjustable injection frequency is generated as shown in Figure 6-12.

To generate injection frequencies separated by 10-kHz steps, the phase comparator operates with a 10-kHz frequency at both inputs. A frequency divider changes the standard frequency to 10 kHz. The programmable frequency divider is controlled by switching circuits. Note that the output from the voltage-

---

[2]A quartz crystal occasionally becomes intermittent and may not oscillate unless the power switch is turned on and off several times. Occasionally, a quartz crystal will suddenly "jump" to another frequency—the crystal must be replaced.

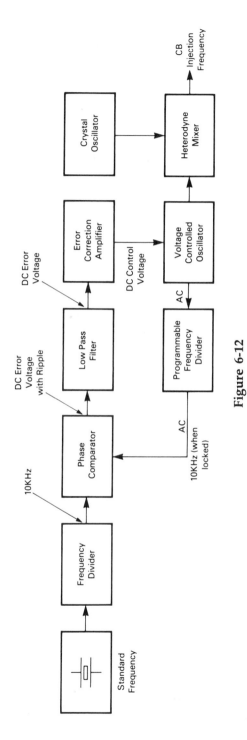

Figure 6-12

Example of a heterodyne frequency synthesizer with phase-locked loop for CB reception.

controlled oscillator (VCO) is in 10-kHz steps, but the least-significant figure for each channel frequency is 5 kHz. Accordingly, another crystal oscillator is included to provide for the 5-kHz factor.

CONCLUSION

Some of the malfunctions that occur in FM/AM receivers and in transceivers are obvious, whereas other malfunctions can be quite difficult to contend with when troubleshooting without service data. When a baffling trouble symptom is encountered, comparison tests should be made if possible. If a normally operating unit is not on hand, it may be possible to borrow one from a coworker, club member, or relative. Sometimes a friend who does not have a normally operating unit will know someone who does, and who is willing to lend it. The bottom line is to follow up on your possibilities, even if the chances might seem remote.

IC FUNCTIONAL OVERVIEW

An integrated-circuit AM/SSB detector arrangement is shown in Figure 6-13. This IC is intended specifically for single-sideband AM receivers. It includes a unity gain amplifier (1) which drives the AM detector. Unfiltered audio output is available, and two separate terminals are brought out for a potentiometer to set the automatic gain control threshold. A variable amplifier (2) amplifies the detector output, controlled by the external threshold potentiometer. Phase correction can be made by another external control through the

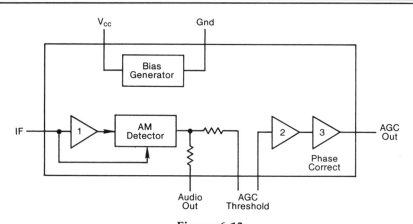

Figure 6-13

Arrangement of an IC AM/SSB detector.

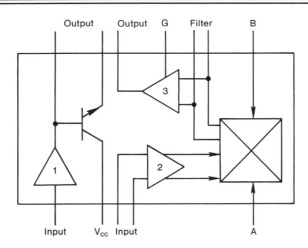

Output   Output   G   Filter   B

3

1

2

Input   V_cc   Input   A

Figure 6-14

Arrangement of an IC multimode detector.

amplifier (3). An internal bias generator circuit provides additional stability over a wide temperature range and variations in V_{cc}. Basic functional parameters are:

1) *Maximum frequency.* The highest rated IF frequency is usually up to 30 MHz. This IC can be operated up to 120 MHz, although some deterioration in performance occurs.

2) *AGC range.* Denotes the input level change required to change the AGC output voltage by 2:1. A typical AGC range value is 5 dB.

This integrated circuit is designed specifically for application in portable or mobile SSB/AM receivers. Note that in some versions of this IC, an audio amplifier following the AM detector is included.

Next, observe the IC multimode detector shown in Figure 6-14. This IC is used as a detector for AM, FM, SSB, and CW signals. Depending on the application, it can be connected to operate as a synchronous detector, quadrature detector, or product detector with a built-in oscillator. For product detection, inputs A and B are used. For FM detection, the input to amplifier (2) is used because this amplifier provides limiting (positive and negative peak clipping).

When limiting is not required, the input to the detector itself can be used. There is a gain-controlled output amplifier (3) with the gain control connected at G. An external filter can be applied at the input to this amplifier. In this example, the IC also contains a separate audio amplifier (1). Basic functional parameters are:

1) *Maximum frequency.* Denotes the highest carrier frequency that the input amplifier (2) will accept. (50 MHz is a typical value.) When amplifier (2) is not used, considerably higher operating frequencies may be employed.

2) *Conversion gain.* This value depends on the setting of the amplifier gain control (G).

   This integrated circuit is used in portable and mobile transceivers of all types. When used as an AM detector, this IC is less noisy and less susceptible to broad-band IF interference than a simple diode detector.

7

TELEVISION TROUBLESHOOTING WITHOUT SERVICE DATA

Overview • Temperature Quick Checks • Typical Causes of Common Trouble Symptoms • Signal-Channel Devices, Frequencies, and Preliminary Test Procedures • Trick of the Trade • IF Oscillation • Case History • High-Voltage Quick Check • Quick Check for DC High Voltage • "Hot Chassis" Caution • AGC Checkout and Troubleshooting • RF AGC • IF AGC • Quick Check for Localization of Intercarrier IF Section • Comparison Quick Check • Spotting AGC Transistors • Integrated Circuits • IC Functional Overview • Sweep Alignment Procedure

OVERVIEW

Some TV receivers are easier to "size up" than others. A few chasses have modular designs wherein each circuit section is fabricated on an individual PC board with plug-in terminals. This type of receiver is comparatively easy to troubleshoot at the modular level. For example, if the trouble symptom is "erratic sync lock," the sync module can be removed and a new sync module plugged in. Of course, the troubleshooter must know where the sync module is located on the chassis, or he or she must make suitable quick checks to identify the sync module.

If a new module is not available, the only alternative is to troubleshoot the defective module. This is a relatively difficult undertaking which usually requires a good knowledge of circuit action and suitable testers to pinpoint the fault. In some cases, the trouble area may be obvious, such as when a transistor "runs a fever," or "runs a chill." (Temperature tests are often speedy guides to circuit-trouble areas.) Overheated resistors are also obvious, and will sometimes eliminate the need for a systematic circuit analysis.

Most TV receivers do not have modular design; however, there are various "landmarks" that can speed up preliminary analysis of circuit sections. The front end, or tuner, is a prominent "module" in any receiver. In turn, the output from the front end always connects to the input of the picture-IF amplifier. The speaker is another prominent "landmark," and it is always connected to the output of the audio amplifier. Again, the volume control is a helpful "landmark" that connects to the output of the sound detector.

A PC board may have numbered component and device numbers; thus, parts may be marked by progressive R, C, D, T, and IC numbers (R = resistors, C = capacitors, D = diodes, T = transistors, and IC = integrated circuits). As illustrated in Figure 7-1, component and device numbers customarily start at the picture-IF amplifier and increase through various sections to the deflection circuitry. Numbers proceed in groups; thus, the first sequence of numbers will relate to the picture-IF parts, the second sequence of numbers will relate to the picture-detector and video-amplifier parts, and so on.

Integrated circuits are often used instead of discrete transistors and resistors; even the smaller TV receivers are likely to have several integrated circuits. As described earlier, an audio-amplifier section may be provided in IC form. We will also find IF-detector-video integrated circuits, sound-if-audio ICs, horizontal-processor ICs that contain horizontal-oscillator, sync circuitry, and

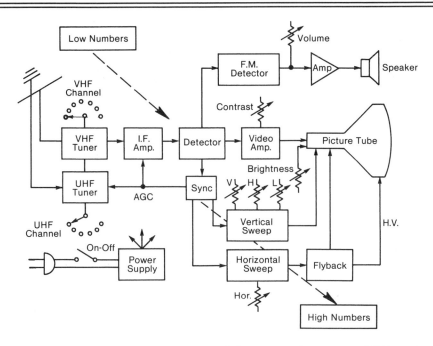

Note: In addition to component and device numbers, some PC boards are also marked with general sectional terminology, such as power, hor, vert, sync, vif, sif, and so on. These markings can be of considerable assistance when troubleshooting without service data.

It is easier to "buzz out" a TV circuit board than a small radio circuit board because a TV board is less densely packed.

Reproduced by permission of Prentice Hall, from Complete TV Servicing Handbook *by Walter H. Buchsbaum.*

Figure 7-1

Low numbers on a PC board generally start with the picture-IF amplifier, and the high numbers generally indicate the deflection circuitry.

horizontal-sweep preamp circuitry, and ICs that process color (chroma) signals. Integrated circuits are prominent devices on a PC board, and their functions can be determined by means of suitable quick checkers.

TYPICAL CAUSES OF COMMON TROUBLE SYMPTOMS

Sound Bars. Display of sound bars on the picture-tube screen points to front-end trouble. If the sound bars cannot be tuned out, and a normal picture tuned in by means of the fine-tuning control, try changing channels. If the reception is normal on other channels, the slug in the oscillator coil for the malfunctioning channel probably needs a touch-up adjustment. On the other hand, when sound bars and/or other abnormal video patterns are obtained on all active channels, it is usually most expedient to replace the tuner.

Hum Output. When black and white (or gray and white) horizontal hum bars are displayed, the trouble is most likely to be found in the power supply. Usually, replacement of the filter capacitors will eliminate the hum bars. Defective filter capacitors can also cause unstable sync lock, hum in the sound, and motorboating (a putt-putt sound output).

Weak and "Silvery" Picture. Picture tube probably needs replacing.

Distorted and Noisy Sound. Ratio detector is likely to be defective; check the electrolytic delay capacitor.

Vertical White Line from Top to Bottom of Screen. Trouble is most likely to be found in the horizontal section of the yoke circuit; an open capacitor is a ready suspect.

Excessively Loud Sound; Little or no Volume Control. The volume-control potentiometer is likely to have developed a defective resistance element.

"Stone-Dead" Receiver. Power-supply trouble is the usual cause.

"Birdies" in the sound. May be accompanied by herringbone patterns in the picture. Picture-IF amplifier malfunction is a prime suspect—open bypass capacitors in the AGC circuit may permit feedback of detector harmonics into the IF channel.

Circuit Breaker Trips Repeatedly. Too much line-current demand; check power consumption with a wattmeter and compare with rated power value on rear of receiver. Excessive current demand may be caused by faulty filter capacitors, or by a defect in a power-transistor circuit. Check operating temperatures of the power transistors with a temperature probe and DVM.

SIGNAL-CHANNEL DEVICES, FREQUENCIES, AND PRELIMINARY TEST PROCEDURES

When a stage or section in the signal channel is "stone dead," although other stages are operational, the operating frequency of the dead stage can be checked with a resonance probe. (The only exception occurs when the fault involves an open or a short-circuited coil winding.)

When checking tuned circuits in a dead stage of a TV receiver, the required generator frequency may be higher, or it may be lower than required in various radio-receiver circuits. High test frequencies for TV circuit tests may be provided by a TV marker generator. Lower test frequencies can be provided by an AM signal generator.

TRICK OF THE TRADE

When "buzzing out" TV circuitry with a mini-amp/speaker (spk), the question may be asked whether a 60-Hz buzz is a video signal or whether it is a vertical-deflection voltage (or a vertical blanking-pulse voltage). This distinction is easily made simply by turning the vertical-hold control up or down. When the vertical-hold control is turned, the vertical-oscillator frequency changes and becomes either higher or lower than 60-Hz. Accordingly, the troubleshooter recognizes *the sound of vertical deflection.*

The audible component of a video signal is the 60-Hz vertical-sync interval; this pitch remains unchanged as the vertical-hold control is turned up or down. Note that when video-amplifier circuitry is checked with a mini-amp/spk in typical receivers, the sound output normally consists of a mixture of the vertical-sync interval and the vertical blanking pulse (if the receiver is tuned to a TV station). Then, when the vertical-hold control is turned up or down, the troubleshooter hears both of the sound sources with their differing pitches.

IF OSCILLATION

In preliminary troubleshooting procedures, it is occasionally observed that the screen is blank, and no picture is displayed when a TV station is tuned in. A video signal cannot be heard at the picture-detector diode with a mini-amp/spk. However, a DC voltage measurement at the detector diode shows that a comparatively high DC voltage is present. In this situation, the troubleshooter concludes that the picture-IF section is oscillating due to some fault or to misalignment.

HIGH-VOLTAGE QUICK CHECK

When the screen is dark, the troubleshooter needs a high-voltage quick check to determine whether there is a fault in the high-voltage section. This is easily determined by means of a miniature neon bulb. If the neon bulb is held in the vicinity of the flyback transformer, the bulb will glow if a normal 15,750-Hz AC field is present. If the neon bulb does not glow, the troubleshooter concludes that the high-voltage section is "dead." (See Figure 7-2.)

A neon bulb consists of a small glass envelope, usually tubular in shape, with a pair of wire leads, or alternatively with a miniature lamp base. In either

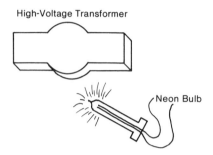

High-Voltage Transformer

Neon Bulb

Note: In normal operation, the neon bulb typically glows at a distance of approximately two inches from the transformer. Observe that the field pattern is very uneven as the bulb is moved in a circle around the transformer.

When a neon bulb is held in the near-field of a high-voltage transformer, the bulb glows if high voltage is present. The striking voltage is approximately 90 V, and the extinction voltage is approximately 60 V. The field has a steep potential gradient, and a 5000-V field potential decreases to 4000-V potential, for example, at a short distance farther away. The electrodes in the neon bulb assume the voltage potentials existing at their positions in space. The high-voltage field alternates 15,750 times per second. As soon as the potential difference between the electrodes rises to 90 V, the bulb glows for a small fraction of a second while the potential difference decays to 60 V. Thus, although the bulb appears to glow continuously, it is actually being energized by narrow pulses of electricity 15,750 times per second.

The neon bulb may be held by its leads, if desired, and no shock will be felt. In other words, the neon bulb typically generates less then 10 V, with a very small current capability, as detailed subsequently. (See also Figures 9-9, 9-10, 9-11, and 10-9.)

Figure 7-2

Miniature neon bulb held in the field of the high-voltage transformer provides a helpful quick check.

case, the bulb contains neon gas at low pressure, with a pair of wire electrodes separated about 1/16 inch. When a voltage of approximately 90 V, or higher, is applied to the leads, the neon gas glows with an orange color. Note that if a neon bulb is directly connected across a voltage source, it will explode because excessive current is drawn, once that the gas has ionized. Therefore, a neon bulb is always connected in series with a resistor, such as 100k, if it is to be connected directly across a voltage source.

If a neon bulb is energized by a DC voltage, only one of the electrodes will glow. (This indicates the polarity of the DC voltage.) However, if the bulb is energized by an AC voltage, both of the electrodes will glow. (This indicates that AC is flowing.) Observe that if the bulb is energized by a pulse voltage, only one electrode will glow. Although a pulse voltage is an AC voltage, it is an unsymmetrical waveform in which one peak voltage is considerably greater than the other peak voltage. (A flyback or high-voltage field represents a pulse voltage.)

Note that when a neon bulb is held in the field of the high-voltage transformer on a TV chassis, the bulb is not actually connected to any point in the high-voltage circuitry. Nevertheless, the bulb glows due to the steep gradient of the field. In other words, the field potential has a high value at the transformer coils, but this field potential decreases rapidly as the bulb is moved away from the coils. It is the potential gradient, or difference in field potential from one electrode to the other electrode that causes the gas to glow. You will observe that if you position the bulb carefully so that the field potential is the same at both electrodes, the bulb does not glow, although the field may be quite strong.

QUICK CHECK FOR DC HIGH VOLTAGE

The presence of AC high voltage can be checked quickly with a neon bulb. However, this test cannot be used to determine whether DC high voltage is present. For example, if the screen is dark, and a neon bulb shows that the high-voltage transformer is working, the troubleshooter next asks whether the high voltage AC is being converted into high voltage DC. This determination can be made quickly by means of a polyester thread, as shown in Figure 7-3. In other words, even if the screen is dark, a polyester thread will be strongly attracted and then strongly repelled as the power switch is turned on and off.

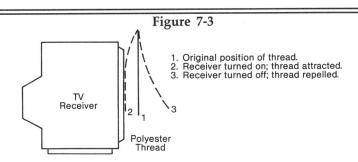

Figure 7-3

1. Original position of thread.
2. Receiver turned on; thread attracted.
3. Receiver turned off; thread repelled.

TV Receiver

Polyester Thread

Note: After the receiver has been turned off for some time, there is no electrostatic charge on the screen. A polyester thread will hang vertically when placed in the vicinity of the screen. Then, when the receiver is turned on, a strong electrostatic field is produced if high

Figure 7-3 (Continued)

DC voltage is present. The thread is attracted in this field and flies over to the screen. In turn, the thread acquires more or less electric charge. Next, when the receiver is turned off, the high DC voltage suddenly ceases, but the thread retains its charge and now flies away from the screen. This is a handy quick check to determine whether high DC voltage is or is not present when troubleshooting a dark screen symptom.

Spools of polyester thread are so-marked on the end of the spool.

Figure 7-3

Quick check for the presence or absence of DC high voltage at the picture tube.

There is a comparatively strong electrostatic field in front of the picture tube in normal operation, and the principle of attraction of unlike polarities and repulsion of like polarities results in the movement of the thread. Next, if there is a fault in the picture-tube circuitry, such as incorrect bias voltage, the screen will be dark. However, if DC high voltage is still present, a polyester thread will indicate its presence when the receiver is switched on and off. After the troubleshooter gains experience with this quick check, he or she can roughly estimate the value of the high DC voltagee.

QUICK CHECK FOR LOCALIZATION OF INTERCARRIER IF SECTION

When "sizing up" a PC board for a TV receiver, the troubleshooter sometimes needs a quick checker for localization of the intercarrier IF section. A handy probe arrangement for this purpose is shown in Figure 7-4. The figure consists of a junction field-effect transistor (JFET) followed by a peak rectifier for indication by a DVM. The JFET provides weak-signal amplification and also develops high input impedance. Since the drain load is tuned to 4.5 MHz, maximum response is obtained in intercarrier circuitry, whereas little or no response is obtained in circuits that operate at other frequencies. In application, the TV receiver must be tuned to a station signal so that an intercarrier IF signal is generated.

In the example shown in Figure 7-4, a fixed base-emitter bias voltage of 3 V is provided. This value is suitable for typical commercial JFETs and operates the transistor in class A. Note however, that some varieties of JFETs will require other bias-voltage values for class-A operation. The rated class-A operating conditions are usually printed on the transistor packet. Troubleshooters generally

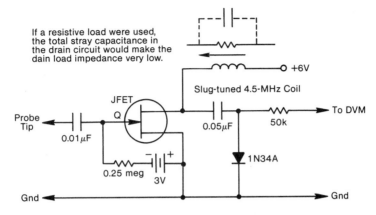

If a resistive load were used,
the total stray capacitance in
the drain circuit would make the
dain load impedance very low.

Slug-tuned 4.5-MHz Coil

JFET

Probe
Tip

0.01 μF

Q

0.05 μF

50k

To DVM

+6V

1N34A

0.25 meg

3V

Gnd

Gnd

Note: This quick-check arrangement identifies intercarrier IF circuitry on the basis of frequency (4.5 MHz). It has comparatively high sensitivity and provides indication even in weak-signal situations. The input impedance to the probe is high so that circuit loading and/or detuning is minimized. Observe that in high-Q circuitry, it is possible that instability might develop (probe produces an output even though there is no input signal). In such a case, the troubleshooter should include a resistor of several thousand ohms in series with the probe tip to stabilize the arrangement. Observe that if a resistive drain load were used in the probe, no identification of 4.5-MHz operation versus other frequencies would be obtained. Moreover, the drain load impedance would be comparatively low.

Figure 7-4

Active probe quick checker for intercarrier section localization and signal tracing.

prefer to use a JFET in probe arrangements, instead of a MOSFET, because the former is a comparatively rugged device that is not easily damaged by accidental overload.

Observe that when the probe is placed on the bench and is not connected into a receiver circuit, it usually picks up stray fields and provides a DVM indication that varies greatly as the probe is moved about. This stray-field response is a result of the high input impedance to the probe, and should not cause concern to the beginner. When the probe is applied at some point in a receiver circuit, the effective input impedance then decreases greatly and takes on the value of the circuit's internal impedance.

If you do not wish to take the time to construct a probe, a trick of the trade can help you quickly localize the general circuit area for the intercarrier sound section on the PC board. Localization is made with the aid of a short-wave AM radio receiver. The receiver is tuned to the intercarrier IF frequency (4.5 MHz), and the end of the antenna rod is used to find the strongest 4.5-MHz field area on the PC board. Note that the TV receiver must be tuned to a station, and the volume control should be turned down.

When the end of the antenna rod on the AM radio is held near the intercarrier sound section on the PC board, the sound output from the radio will be quite loud and clear. On the other hand, in the vicinity of the video amplifier or other receiver sections, the sound output from the radio will be comparatively weak, noisy, and often distorted. In normal operation, the signal-to-noise ratio is quite good near the intercarrier sound circuitry.

This quick check is based on the circumstance that although the intercarrier-IF signal is frequency modulated, there is a substantial amount of incidental amplitude modulation (IAM) generated by slope detection in the intercarrier signal channel. For example, the sound carrier in the picture-IF section falls on the lower slope of the IF frequency-response curve. IAM is usually regarded as only a spurious signal component to be eliminated by limiter action and ratio detection. However, IAM can serve a useful purpose in preliminary troubleshooting procedures as explained above.

INTEGRATED CIRCUITS

Integrated circuits are found in many TV receivers. Audio-amplifier ICs were discussed previously in Chapter 3. In addition to these ICs, troubleshooters will encounter integrated video amplifiers, integrated TV sound circuits comprising FM amplifier/limiter, FM demodulator, and audio preamplifier/driver sections, integrated IF amplifiers, and various arrangements such as those shown in Figure 7-5. Observe that this IC comprises an IF section, detector, video amplifier, AGC, and automatic fine-tuning (AFT) circuitry.

In each case, an integrated circuit contains only the active devices and some of the resistors that are used in the complete configuration. Thus, inductors, tuning capacitors, fixed capacitors, and potentiometers are external to the IC. When trouble symptoms develop, the first question is whether the fault is in the IC, or in some external component. Accordingly, the troubleshooter makes tests of the external components that could be involved in the malfunction. Then, if it appears that the external components are not defective, the troubleshooter proceeds to replace the integrated circuit.

Note that troubleshooting of the more elaborate black-and-white TV receivers without service data can be a formidable challenge unless the trouble symptom is clear cut. Accordingly, it is often impractical to troubleshoot an ambiguous trouble symptom except on a comparison basis with a similar receiver that is in normal operating condition.

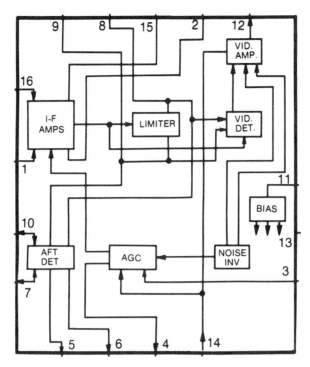

Note: This is an example of an IC package that contains a large number of built-in transistors, diodes, and resistors. However, the troubleshooter is not concerned with the internal circuitry. If the arrangement does not work, and the external components appear to be ok, the troubleshooter just replaces the IC. Observe that the DC supply voltages must be normal, and the signal input voltages must be normal before it is concluded that the IC is defective.

In the absence of service data, these values may be obtained by comparative measurements on a normally operating receiver.

Reproduced by permission of Prentice Hall, from Complete TV Servicing Handbook, *by Walter H. Buchsbaum.*

Figure 7-5

Example of an integrated circuit that provides several TV signal-processing functions.

An example of a detector and audio section in integrated-circuit form for a TV receiver is shown in Figure 7-6. It employs DC volume control. Practically, this means *the troubleshooter will not find any audio signal voltage at the volume control.* Here, the volume control is actually a DC-voltage bias control. It varies the bias on an amplifier network inside of the IC, thereby determining the level of audio signal flow.

IC FUNCTIONAL OVERVIEW

The functional arrangement of a widely used integrated circuit for automatic fine-tuning (AFT) is shown in Figure 7-7. This IC operates from the video-IF input; it contains an internal shunt regulator and a bias generator. In addition to the AFT voltage, it also generates an amplified 4.5-MHz intercarrier sound signal. The AFT circuit employs a differential peak detector with amplifier, and

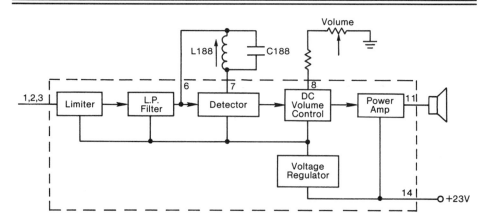

Note: This is an example of an intercarrier FM detector and audio amplifier in IC form. Observe that a DC volume control is used in this arrangement. Functionally, a DC volume control varies the bias voltage on a differential-amplifier network inside of the integrated circuit, thereby varying the audio signal level. Observe that the external tuned circuit operates at 4.5 MHz.

Reproduced by permission of Prentice Hall, from Complete TV Servicing Handbook, *by Walter H. Buchsbaum.*

Figure 7-6

Example of an audio section integrated circuit that employs DC volume control.

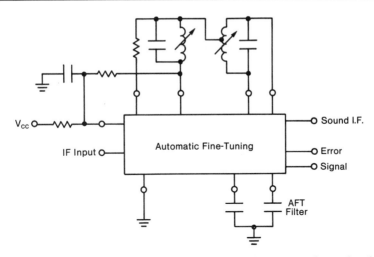

Technical Note: Automatic fine-tuning troubles are evaluated to best advantage by means of comparison tests with respect to a similar receiver that is in normal operating condition. The pull-in range is a key consideration; observe also the change in tuning characteristics when the AFC defeat switch is thrown. Comparative DC voltage and resistance measurements will often localize the trouble area, if not the faulty device or component.

Figure 7-7

Basic functional arrangement of an automatic fine-tuning (AFT) integrated circuit.

provides bipolar positive and negative error signals to the TV tuner. The time-constants of the AFT circuit are determined by an external, two-capacitor AFT filter, as shown in the diagram. Basic functional parameters for this IC are:

1) *Signal input.* Denotes the amplitude of the IF input that is required (15 rms mV is a typical rating).

2) *Error signal amplitude.* This value depends on the deviation of the IF frequency from its desired value. When the IF frequency is 44.65 MHz, the error voltage will be, typically, between 2.2 and 4.7 V. When the IF frequency is 46.85 MHz, the error voltage will vary from 9.1 to 12.1 V. At the terminal of the opposite polarity, the error voltage will be opposite, in relation to the IF frequency.

3) *4.5 MHz output voltage.* With an IF input signal consisting of 45.75 MHz

and 41.25 MHz carriers, the 4.5-MHz output signal will be approximately 11 rms mV.

In typical applications, the error signal is passed through series resistors and shunt bypass capacitors and is applied across a varactor diode. A varactor diode changes its junction capacitance in response to changes of voltage across its terminals. In turn, the varactor diode is a functional part of the local oscillator resonant circuitry in the TV tuner, and thereby determines the precise operating frequency of the local oscillator.

SWEEP-ALIGNMENT PROCEDURE

Most troubleshooters prefer a sweep-alignment procedure whenever an extensive tuned-circuit adjustment is required in a TV receiver. The principles involved, and the practical test set-ups are shown in Figure 7-8. The sweep generator applies a large-deviation frequency-modulated signal to the tuned circuitry for display of its amplitude-versus-frequency characteristic. This sweep signal has a repetition rate of 60 Hz.

Figure 7-8

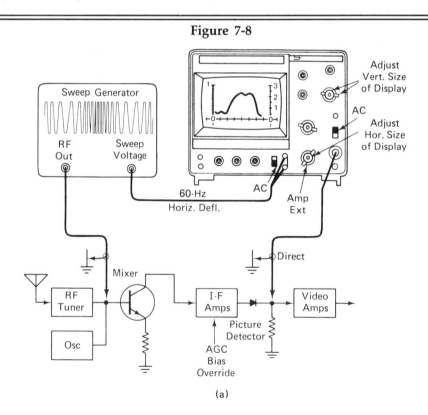

(a)

Figure 7-8 (Continued)

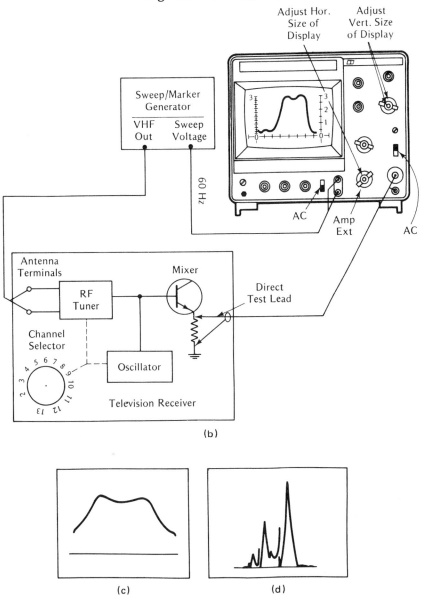

(b)

(c)

(d)

Figure 7-8 (Continued)

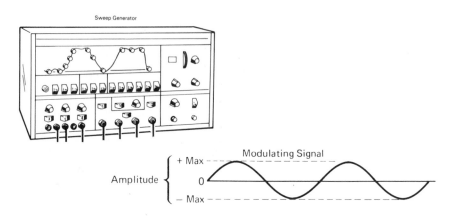

Sweep Generator

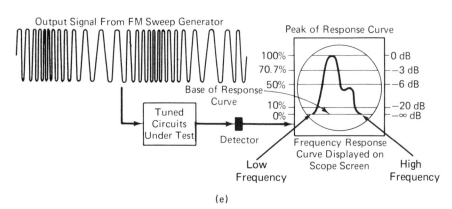

Modulating Signal

Amplitude {
+ Max
0
− Max

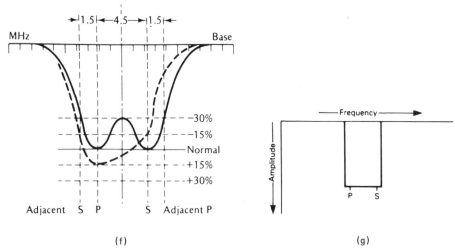

Output Signal From FM Sweep Generator

Peak of Response Curve

100% — 0 dB
70.7% — −3 dB
50% — −6 dB

Base of Response Curve

10% — −20 dB
0% — −∞ dB

Tuned Circuits Under Test

Detector

Frequency Response Curve Displayed on Scope Screen

Low Frequency   High Frequency

(e)

MHz   Base

→|1.5|←→4.5→|1.5|←

−30%
−15%
Normal
+15%
+30%

Adjacent   S   P   S   Adjacent P

(f)

Frequency →

Amplitude

P   S

(g)

Figure 7-8 (Continued)

Frequency Reference Chart

Channel Number	Picture Carrier Frequency (MHz)	Sound Carrier Frequency (MHz)	Receiver VHF Oscillator Frequency (MHz)
2	55.25	59.75	101
3	61.25	65.75	107
4	67.25	71.75	113
5	77.25	81.75	123
6	83.25	87.75	129
7	175.25	179.75	221
8	181.25	185.75	227
9	187.25	191.75	233
10	193.25	197.75	239
11	199.25	203.75	245
12	205.25	209.75	251
13	211.25	215.75	257

(h)

Courtesy, B&K PRECISION, Div. of Dynascan Corp.

Figure 7-8

Sweep-alignment principles and test setups: (a) IF alignment arrangement; (b) front-end alignment arrangement; (c) normal VHF response curve; (d) typical distorted VHF response curve (regenerative); (e) dB points on a response curve; (f) permissible tolerances; (h) frequency reference chart.

The oscilloscope is connected at the output of the detector following the tuned circuits. This detector might be a heterodyne mixer or converter, a second detector, or a ratio detector. In any case, the detector demodulates the sweep voltage, and a demodulated frequency-response curve is displayed on the scope screen. Most sweep-alignment generators have built-in marker generators. Alternately, the troubleshooter can inject a small marker voltage into the swept circuitry from a separate marker generator.

ADDITIONAL TELEVISION TROUBLESHOOTING METHODS

Signal-Killer Quick Checker • Integrated Circuit Identification • Continuity Testing on PC Boards • Marked Jumpers • Subsection Parts Numbering • Typical PC Board Layout • How Hot? • Separation of Sound and Picture • Impedance Checks in Sync Circuitry • Troubleshooting Power Supplies • IC Functional Overview

SIGNAL-KILLER QUICK CHECKER

When troubleshooting television receivers without service data, questions often occur concerning the location of various functional sections on the PC board. If the receiver is workable, although malfunctioning in some regard, the troubleshooter can often speed up localization of functional sections by means of the signal-killer quick checker shown in Figure 8-1. The signal killer consists of a 0.1 μF capacitor in series with a clip lead. It is used to short-circuit the signal at a test point to ground, and the receiver picture and sound channels are used as indicators.

For example, if the troubleshooter needs to locate the general area of the sync section on the circuit board, he or she proceeds to short the signal at "likely" terminal points to ground. When the troubleshooter "hits" a horizontal-sync signal point, the picture will lose horizontal sync lock. (Or, if horizontal sync action is poor before the test is made, sync action will become poorer during the test.) The particular change that takes place will depend on the particular circuit branch that is being tested. However, the essential consideration is that the troubleshooter can determine that he or she is checking in the horizontal-sync section on the basis of reaction in the picture display.

It is evident that a signal-killer quick check in the video-amplifier section will "wipe out" the picture, although the sound output will be unaffected. Or, if the signal is killed in the vertical-oscillator section, only a white horizontal line will be displayed on the picture-tube screen. Again, if the signal is killed in the horizontal-oscillator section, the screen will become dark, due to loss of high voltage. Observe that if the signal is killed in the sync-separator section, both horizontal sync lock and vertical sync lock will be lost. If the signal is killed in the vertical-blanking section, vertical retrace lines will appear in the picture. In the case of keyed AGC, if the keyer signal is killed, the picture contrast will increase excessively.

INTEGRATED CIRCUIT IDENTIFICATION

When an unfamiliar PC board is being "sized up," the question of identification of integrated-circuit packages frequently arises. Identification can be achieved easily by means of signal-killer quick checks, as shown in Figure 8-2. The troubleshooter merely "wipes" the pins on the IC package with the signal-killer test lead, and observes the resultant picture-and-sound response. In this figure, a video-IF IC package is being tested, and the signal killer is contacting

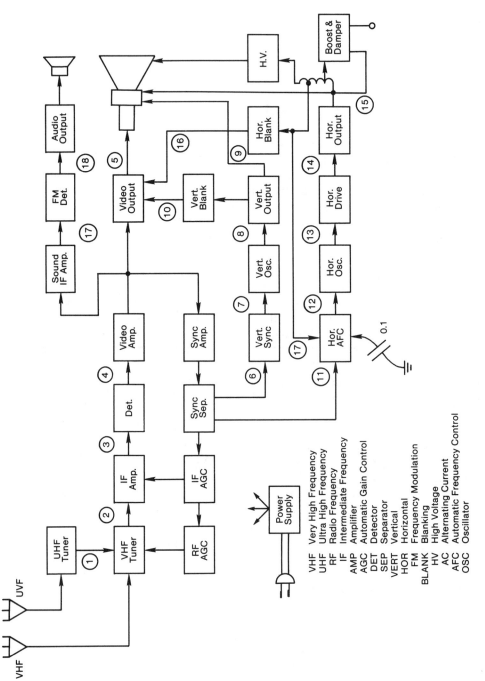

Figure 8-1

Figure 8-1 (Continued)

Note: This localization procedure uses a 0.1μF fixed capacitor with a clip lead to short-circuit various terminal points on the PC board to ground (AC short-circuit). In turn, the picture display and sound output responses are evaluated for clues to identification of the terminal point under test. Thus if the horizontal-AFC signal is shorted to ground, horizontal sync lock will be lost. Or, if the vertical sync signal is shorted to ground, vertical sync lock will be lost. Again, if the sound-IF signal is shorted to ground, sound output will stop.

Caution: Avoid short-circuiting terminal points in the high-voltage and horizontal-output sections to AC ground—possible overload damage could occur.

Reproduced by permission of Prentice Hall from Complete TV Servicing Handbook, *by Walter H. Buchsbaum.*

Figure 8-1

Example of functional section localization by means of signal-killer quick checks.

the video-output terminal. It is evident that when the video signal (with its accompanying intercarrier-sound signal) is killed, the picture-tube screen will go blank and the speaker will be silent.

Of course, if the signal channels in the receiver happen to be "dead," then a signal-killer quick check cannot be made. As noted previously, a TV IF resonance probe is the best quick-checker to use in this situation. Observe that in the example of Figure 8-2, T1 and T2 are available for a resonance-probe quick check. The troubleshooter can therefore establish the function of the IC, although the signal circuits are "dead."

CONTINUITY TESTING ON PC BOARDS

When "sizing up" or "buzzing out" a PC board, it is often helpful to make continuity quick-checks between various terminals, instead of visually following the course of a comparatively long "trace." To make effective continuity quick checks, a conventional continuity tester should not be used, inasmuch as indications may be false. In other words, the troubleshooter needs to know whether a conductive path is present between the points under test. A

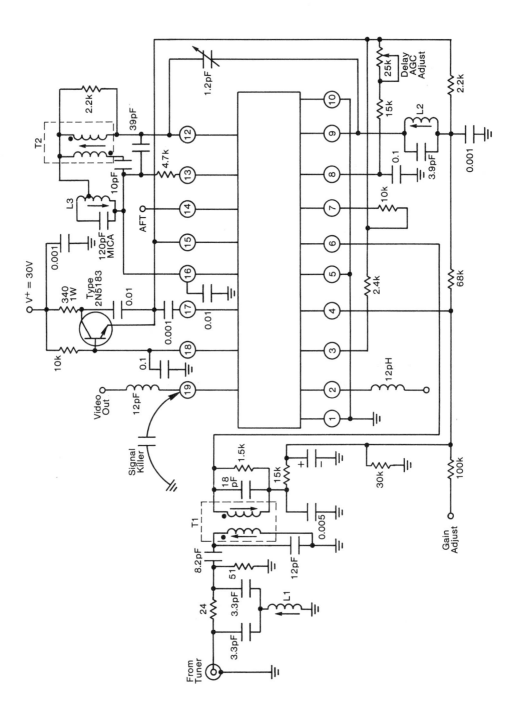

Figure 8-2

Figure 8-2 (Continued)

Note: In this method of integrated-circuit identification, the signal at various pins is killed and the resultant response is observed on the picture-tube screen and from the speaker. Thus, a picture-IF integrated circuit is being quick-checked in this example. The signal killer happens to be applied at the video output terminal of the IC. In turn, both the picture and the sound will be killed. Observe that even if the IC (or its associated external components) are faulty, but more or less picture and/or sound is present, the quick check is still effective in indentifying the IF integrated circuit.

Reproduced by permission of Prentice Hall from Complete TV Servicing Handbook, *by Walter H. Buchsbaum.*

Figure 8-2

Example of identification of an integrated circuit.

conductive path presents almost zero resistance, whereas a commercial continuity tester typically indicates continuity for resistance up to 600 ohms, or a semiconductor junction.

Therefore, the troubleshooter should always use a low-power ohmmeter to make continuity tests on PC boards. The low-power ohmmeter will show whether the path between the points under test is practically zero, or whether it may have some finite resistance value. Moreover, a low-power ohmmeter cannot "turn on" normal semiconductor junctions. The only disadvantage of using an ohmmeter in this application is the necessity of repeatedly glancing at the display instead of listening for a "BEEP."

MARKED JUMPERS

TV PC boards may have jumpers connecting two terminals on the solder side of the board. A jumper is similar to a hairpin; it is inserted from the component side of the board, and is soldered to pads on the solder side of the board. If components and devices are marked with numbers, jumpers will also be marked. From the troubleshooter's viewpoint, jumpers are sometimes marked for their normal DC voltage (and perhaps current). This can be quite helpful on occasion. As an illustration, a typical marked jumper in the power-supply section of a popular TV receiver is marked "106 V, 4 mA." Accordingly, if power-supply malfunction is suspected, the troubleshooter can make a quick voltage and/or current measurement to verify or dismiss the suspicion. (See Figure 8-3.)

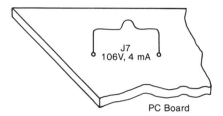

PC Board

Note: A jumper is a U-shaped wire connector inserted into a pair of eyelets. When marked with a rated voltage value, this voltage is normally measured from the jumper to ground (common bus). Since the jumper is soldered in place beneath the PC board, the troubleshooter may opt to measure the current by cutting the jumper wire, and connecting a milliammeter to the ends. After the current measurement is completed, the jumper can be repaired with a small drop of solder.

If marked jumpers are provided on a PC board, the transistor eyelets may also be marked E, B, and C. These markings can save considerable time in preliminary troubleshooting procedures for measurement of bias and collector voltages.

In some cases, it will be observed that marked PC boards have various "missing" components and devices. This is because the same PC board is used in different models of the general type of receiver. The "missing" parts are disregarded in "sizing up" and "buzzing out" the PC board.

Figure 8-3

Jumpers may be provided, with DC voltage and current markings.

SUBSECTION PARTS NUMBERING

In some cases, a functional section may be separated into two subsections on the PC board, as exemplified in Figure 8-4. Here, the subsections are far apart; however, the troubleshooter can recognize them as part of the same section because the same group of numbers are used for the parts numbers in both subsections. Accordingly, this is a helpful point to keep in mind when "sizing up" a PC board. Observe that the conductive path from the sound-IF subsection to the audio subsection (curved dotted line) is not a part of the

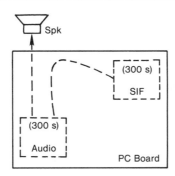

Note: In this example of separated subsections, the sound-IF (SIF) section is located in the upper right-hand corner of the PC board, whereas the following audio section is located in the lower left-hand corner of the board. Observe that since both of these subsections are part of the sound section, the same group of numbers are used in the separated subsections (the 300 series of numbers, in this example). Note that the volume control (not shown) is located near the speaker.

Figure 8-4

Subsections are occasionally separated on the PC board.

printed circuitry. Instead, it is a coaxial cable that connects to the volume control, and then to the audio subsection.

TYPICAL PC BOARD LAYOUT

It is helpful for the troubleshooter to keep typical PC board layouts in mind when "sizing up" and "buzzing out" a TV receiver. As an illustration, a typical sectional subsectional layout for the PC board of a small TV receiver is shown in Figure 8-5. Observe that the video-IF, AGC, and video-amplifier comprise three subsections, and that 100-series parts numbers are used in all three subsections. In this example, the common practice of locating the AGC level and delay controls in the AGC subsection is utilized. (The brightness and contrast picture controls are mounted on the rear of the PC board.)

Sync controls are also mounted on the rear of the PC board, and the volume control is mounted above the speaker in the figure. Also, the common practice of locating the vertical linearity and height controls in the vertical-deflection section is utilized. A transformerless power supply is employed with an 800-mA line fuse. A 145° thermal fuse is also provided in the horizontal-

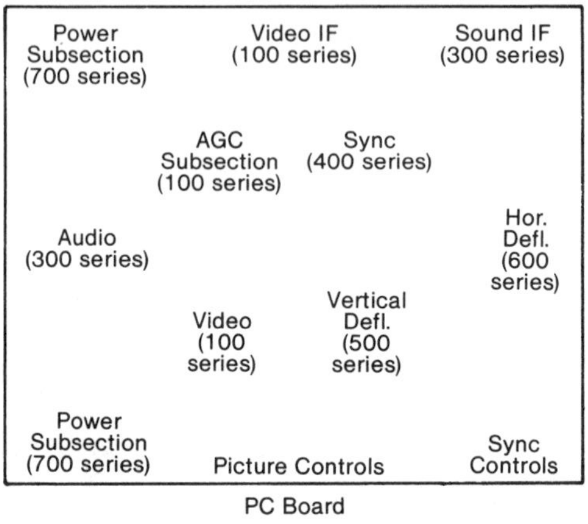

Power Subsection (700 series)	Video IF (100 series)	Sound IF (300 series)

AGC Subsection (100 series)   Sync (400 series)

Audio (300 series)   Hor. Defl. (600 series)

Video (100 series)   Vertical Defl. (500 series)

Power Subsection (700 series)   Picture Controls   Sync Controls

PC Board

Note: Small-signal transistors, power-type transistors, and integrated circuits are used in various sections and subsections. The power-type transistors are located easily by means of their heat sinks (vertical L-shaped sheet-metal units). A power transistor is secured to the side of each heat sink (four heat sinks are provided in this example). Observe that the sound-IF and audio groupings are subsections in this arrangement, and 300-series parts numbers are used in both subsections.

Figure 8-5

Typical sectional and subsectional PC board layout for a small black-and-white TV receiver.

deflection section, in this example. As noted previously, troubleshooting requires the use of a line-isolation transformer to avoid "hot chassis" shock hazard, and also possible damage to test equipment.

HOW HOT?

When troubleshooting TV receivers without service data, it is helpful to have a general idea of how hot various power components and devices run in normal operation. As a general guideline, the following temperatures may be noted: ambient, 21°C; horizontal-output transistor, 30°C; vertical-output integrated circuit, 42°C; audio transistor, 57°C; power-supply transistor, 47°C; thermal fuse, 90°C (blows at 145°); line fuse, 26°C; 10-watt power resistor, 150°C.

The video amplifier output transistor in this guideline example is larger than the small-signal transistors, but it is not mounted on a heat sink. Its normal operating temperature is 33°C. Note that these are "ballpark" temperatures that will vary somewhat from one receiver to another. However, these temperatures provide a useful guide concerning whether a power component or device is "running a fever," or "running a chill."

SEPARATION OF SOUND AND PICTURE

When the sound reproduction is best at one position of the fine-tuning control, but the picture is poor or invisible, or the picture reproduction is obtained at another position of the fine-tuning control, but the sound is poor or inaudible, the troubleshooter describes the symptom as "separation of sound and picture." In this situation, the picture is objectionably distorted, even at the optimum setting of the fine-tuning control. The separation trouble symptom is almost certain to be associated with subnormal bandwidth in the IF section, due to spurious positive feedback. The most likely culprit is an open bypass or decoupling capacitor.

Separation of sound and picture is often accompanied by unstable sync lock. Complete loss of sync may occur at some setting of the fine-tuning control. A trick of the trade that is helpful in preliminary analysis is to place a screwdriver in the vicinity of an IF transformer. In case spurious positive feedback occurs, the picture will be highly responsive to movement of the screwdriver. Note that in exceptional cases, separation of sound and picture may be caused by spurious feedback in the tuner.

TROUBLESHOOTING POWER SUPPLIES

Power supplies in TV receivers are often regulated, and it is helpful for the troubleshooter to recognize the two basic types of regulator circuits and their principles of operation.

Series regulator circuits such as those shown in Figure 8-6 employ direct-coupled amplifiers to step up an error (difference) voltage, which is obtained from a comparison between a portion of the output voltage and a reference source. This reference source is typically a zener diode. Observe that the reference-voltage source V_R is connected in the emitter branch of the amplifier transistor Q_1, so that the error or difference voltage between V_R and some portion of the output voltage V_0 is applied between base and emitter, and in turn, is amplified.

The amplified error voltage is the input to the regulating element; this regulating element comprises transistors Q_2 and Q_3, and the output voltage from the regulating element is the controlling voltage that appears across R_1. Observe that the current flow through Q_3 produces an IR drop across R_2 and R_3, which is the regulated output voltage V_0. Thus, Q_3 is in series with the input voltage and the regulated output voltage. Note that if the input voltage in-

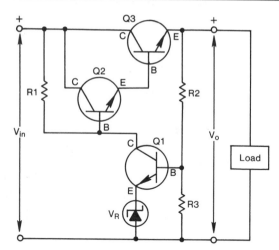

Note: This series-regulator arrangement is in very wide use. It is called a series regulator because transistor Q_3 operates in series between V_{in} and V_0. The output voltage V_0 is always less than the input emf V_{in}. Thus, the difference between the input and output voltages is dropped from collector to emitter of Q_3. Functionally, Q_3 "looks like" a variable resistor that increases its value if V_{in} starts to rise, or decreases its value if V_{in} starts to fall. Similarly, Q_3 decreases its resistance value if V_0 starts to fall as a result of heavier loading (increased current demand). Or, Q_3 increases its resistance value if V_0 starts to rise as a result of lighter loading (decreased current demand). Observe that the portion of V_0 that is dropped across R_3 serves as a bias voltage to effectively control the collector resistance value of Q_3.

Figure 8-6

The basic series regulator configuration.

creases, the error voltage starts to increase. Consequently, Q_1 draws more current, and the base-emitter bias on Q_2 decreases. Q_2 draws less current, and Q_3 draws less current. Therefore, less current now flows through R_2 and R_3 and V_0 is maintained near its rated value.

Next, suppose that the load on the power supply is increased; the current demand increases accordingly, and the value of V_0 would decrease if the power supply were unregulated. Note that when V_0 starts to decrease, the error voltage starts to decrease. Accordingly, Q_1 draws less current, and the base-emitter bias on Q_2 increases. Thus, Q_2 draws more current, and Q_3 draws more current. Therefore, more current now flows through R_2 and R_3 and V_0 is maintained near its rated value.

Observe that Q_2 and Q_3 effectively comprise a Darlington emitter follower with respect to the output resistors R_2 and R_3. We will recall that the output impedance of a Darlington emitter follower is very low. This is just another way of saying that the output impedance of the regulated power supply is very low (much lower than the resistance of $R_2 + R_3$). Note too, that the regulator circuit can "follow" changes of input voltage rapidly. For this reason, the voltage-regulator circuit also has substantial filtering action, and largely smoothes out any ripple that may be present on the input voltage.

The "acid test" of a regulator arrangement is its ability to maintain a constant output voltage as the line voltage is varied ± 10 percent, for example, and as the load-current is varied ± 10 percent.

Shunt regulator circuits such as those shown in Figure 8-7 also employ direct-coupled amplifiers to step up an error (difference) voltage, which is obtained from a comparison between a portion of the output voltage and a reference source such as a zener diode. Observe that the error voltage is applied as a bias voltage between the base of Q_2 and ground. In turn, the base currents of Q_2 and Q_1 are amplified, with resultant change in the collector-to-emitter resistances of Q_2 and Q_1. If the load-current demand decreases, V_0 tends to rise. However, the collector-current demands of Q_2 and Q_1 then increase, with the result that V_0 is maintained near its rated value.

Figure 8-7

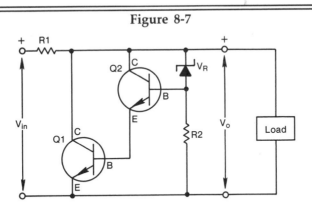

Note: The troubleshooter occasionally encounters shunt regulator arrangements. A shunt regulator configuration is so-called because any current not demanded by the load is shunted to ground via Q_2 and Q_1. Observe that if the source emf V_{in} starts to increase, the output voltage V_0 will also start to increase. However, the current through V_R and R_2 also increases, with the result that the bias voltage at B increases. Accordingly, the collector-to-ground resistance of Q_1 and Q_2 decreases, and more current is drawn via Q_2 and Q_1 to ground. In turn, the voltage drop across R_1 increases, and V_0 is maintained near

Figure 8-7 (Continued)

its rated value. The shunt regulator arrangement is comparatively inefficient because Q2 and Q1 dissipate power, and R_1 also dissipates power in addition to the power demand of the load.

Figure 8-7

Basic shunt regulator configuration.

As in the case of the series-regulator configuration, the shunt-regulator circuit can "follow" changes of input voltage rapidly. Accordingly, ripple that may be present on the input voltage is largely smoothed out. Thus, the regulator circuit functions also as an effective filter. From the troubleshooter's viewpoint, the "acid test" of the regulator secton is its ability to maintain a constant output voltage as the line voltage is varied ± 10 percent, for example, and as the load-current demand is varied ± 10 percent.

IC FUNCTIONAL OVERVIEW

A widely used sound-IF and audio output integrated circuit is shown in Figure 8-8. The IC accepts the 4.5-MHz sound subcarrier signal through a transformer or other resonant circuit; it provides all of the limiting, FM detection, and amplifying functions. As seen in the diagram, the FM detector employs a resonant circuit which is tuned to the 4.5-MHz center frequency. The volume control is a linear taper DC control that connects to an internal electronic attenuator circuit.

This circuit provides an improved audio taper response and drives the audio power amplifier in conjunction with the deemphasis network. Since this IC provides a significant amount of audio output power, a voltage regulator is included with a thermal and current sensing shutdown circuit. Note that the IC has a copper strap heat sink designed as part of the IC body. Basic functional parameters for the IC are:

1) *Maximum power dissipation.* Without an external heat sink, up to 25°C, 1.4 W is maximum. With a copper-strap heat sink soldered to the IC board, up to 25°C, 3.9 W can be dissipated as maximum.

2) *IF input amplitude for limiting.* Amplitude of the 4.5-MHz IF signal that will cause 3-dB limiting is typically 200 microvolts.

3) *AM rejection.* Denotes the ability of the FM detector to reject amplitude modulation. A value of 50 dB is typical.

4) *Total system harmonic distortion.* The total harmonic distortion (THD) from the IF input to the audio output is typically 1.5 percent for a 1-W audio output level.

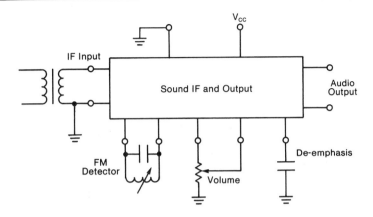

Technical Note: Preliminary troubleshooting procedure can be made to advantage with signal present, and on the basis of comparison tests with an oscilloscope (preferably dual trace). This signal-tracing procedure usually suffices to localize the malfunctioning section of the active network, and follow-up DC voltage and resistance measurements are then usually employed to identify the defective device or component.

Figure 8-8

Functional overview of an integrated circuit for processing the sound IF and output signal.

5) *Power output.* Denotes the maximum audio power output with a total harmonic system distortion of 10 percent. Ratings range from 1 up to 5 W, when adequate heat sinks are used.

Note that exceeding the temperature limits or maximum power dissipation ratings will result in automatic shutdown of the audio section. Later, when the temperature of the equipment has lowered, normal operation will be restored. (This thermal action can be mistaken for an intermittent circuit fault.)

9

PROGRESSIVE TELEVISION TROUBLESHOOTING METHODS

"Buzzing Out" Circuits from Component Side of PC Board • "Garbage" in the V_{cc} Line • Intermittent Monitoring • Constant-Current Impedance Quick Checker • Vertical Sync, Oscillator, and Deflection System • Noise Minimization • Pointers on Circuit Action • Biased Neon-Bulb Quick Checker • Experiment with DC Voltage Generated by a Neon Bulb

"BUZZING OUT" CIRCUITS FROM THE COMPONENT SIDE OF THE PC BOARD

Considerable time can be saved in various situations by "buzzing out" a circuit from the component side of the PC board, as shown in Figure 9-1. A low-power ohmmeter is used to determine whether there is a PC conductor running between a selected pair of test points. PC connections among transistors, resistors, capacitors, and integrated circuits can usually be determined in this

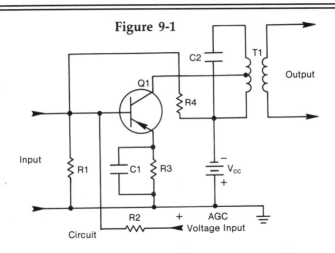

Figure 9-1

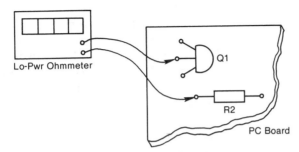

Note: Most circuit tracing can be accomplished from the component side of the PC board with a low-power ohmmeter. This is a helpful procedure in various troubleshooting situations because it eliminates

Figure 9-1 (Continued)

the necessity of "flipping" the board over and over. Observe in the diagrams that the base of Q_1 is connected to R_2. In turn, there is a section of conductor under the PC board that makes this connection. When a low-power ohmmeter is applied between the base terminal of Q_1 and the appropriate end of R_1 on the component side of the board, the ohmmeter will read practically zero.

Note that short lengths of transistor leads are usually accessible on the component side of a PC board. However, if a transistor lead is encountered which is "hidden" under the transistor, the troubleshooter then has no choice but to "flip" the board over.

Figure 9-1

"Buzzing out" a circuit from the component side of the PC board.

manner without difficulty. One exception is tuned transformers, such as T1. The transformer is usually placed directly on top of the PC board without any clearance for a test prod. Accordingly, the board must be "flipped over" in such a case, and the "trace" followed by eye.

Sometimes the troubleshooter needs to establish only one or two interconnections. In turn, a permanent record of the test is not required. On the other hand, the troubleshooter may need to "buzz out" a complete circuit section. In this situation, it is difficult or impossible to remember all of the details of interconnections. There are two practical ways to keep a record of the tests for future reference. The easiest and most direct way is to use a felt-tipped pen, and to draw the interconnection on the board as soon as it is determined. Accordingly, the troubleshooter quickly develops a "circuit diagram" on the PC board.

Another way of keeping a record of the tests is to sketch the components of interest on a sheet of paper, *using schematic symbols.* In turn, the interconnections can be added one by one to the schematic symbols as they are "buzzed out." From the busy troubleshooter's viewpoint, it is generally more desirable to draw the "circuit diagram" directly on the PC board. If the base, emitter, and collector terminals of the transistors are not marked on the board, it can be helpful for the troubleshooter to also letter the transistor terminals for quick identification at a later time.

The advantage of sketching the "buzzed out" circuit on a sheet of paper, using schematic symbols, is that the completed diagram is much easier to analyze than the lines drawn on the PC board. Analysis is required when the cause of the trouble symptom is not obvious. A comparison between a PC board "diagram" and its corresponding schematic diagram is exemplified in Figure 9-2. Observe that it is much more difficult to "follow" the PC board "diagram," than to read the schematic diagram.

"GARBAGE" IN THE V_{cc} LINE

As shown in Figure 9-3, the V_{cc} line will eventually develop excessive ripple and "garbage" due to deteriorating electrolytic capacitors. This V_{cc} inter-

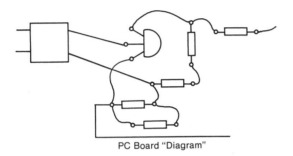

PC Board "Diagram"

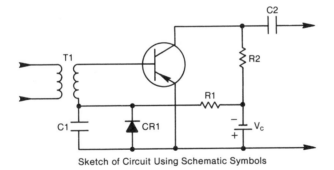

Sketch of Circuit Using Schematic Symbols

Note: It is evident from this example that a PC board "diagram" is comparatively difficult to analyze. Even if the troubleshooter marks the components and devices, evaluation of the arrangement can be confusing. However, if the circuit is sketched on paper with standard schematic symbols, the evaluation and analysis task becomes comparatively easy.

Figure 9-2

Example of PC board "diagram" and corresponding sketch of circuit with schematic symbols.

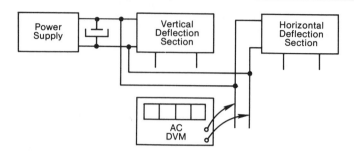

Note: The output voltage from the power supply is filtered by means of a large electrolytic capacitor. In addition, the V_{cc} supply lines to the vertical and horizontal deflection sections are often decoupled by other electrolytic capacitors. As the capacitors age, the "garbage" on the V_{cc} line increases. Ripple from the power supply has a frequency of 60 or 120 Hz and has a typical semi-sawtooth waveform. "Garbage" from the vertical deflection section often has a distorted pulse waveshape with a 60-Hz repetition rate. "Garbage" from the horizontal deflection section has a frequency of 15,750 Hz, with a semi-sawtooth waveform.

Figure 9-3

Puzzling sync symptoms can result from power-supply and deflection circuit interference in the V_{cc} voltage.

ference can enter the sync section via the collector and bias supply lines. The result is that sync action becomes impaired, and the cause may go entirely unsuspected unless the troubleshooter checks the V_{cc} line with an AC DVM or oscilloscope.

INTERMITTENT MONITORING

The most difficult "tough-dog" sync troubleshooting problems occur when the trouble symptoms are intermittent. A long wait may ensue before the trouble symptom "shows up," and it will sometimes disappear before the troubleshooter has sufficient time to check out the fault possibilities. In such a case, the intermittent monitoring procedures noted previously for radio receivers should be used. The essential questions are the same for either radio or TV receivers: How do various DC voltages respond when the intermittent symptoms occur, how do signal voltages respond, or, in some situations, how do current values change when the intermittent symptom occurs?

Observe that faulty horizontal sync lock can be caused by defects in

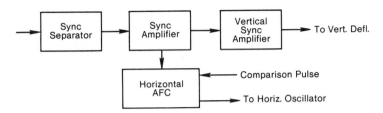

Note: The output from the sync separator consists of both horizontal sync pulses and vertical sync pulses. In turn, the sync amplifier has two outputs; one output goes through a low-pass filter into the vertical sync amplifier, and the other output goes through a high-pass filter into the horizontal AFC section. Here, the timing of the horizontal sync pulse is compared with the timing of a comparison pulse from the horizontal oscillator. Finally, a DC control voltage is outputted from the AFC section to the horizontal oscillator section. This DC control voltage is an error voltage that functions to bring the horizontal oscillator into precise timing with the sync pulse.

Technical Note: Troubleshooting of the sync section generally starts with scope waveform tests (preferably on a comparison basis, using a dual-trace scope). After the malfunctioning subsection has been localized, transistors can be checked with an in-circuit transistor tester. Follow-up DC voltage and resistance measurements will usually serve to close in on the defective component.

Figure 9-4

Relationship of the horizontal-AFC section to the sync section.

circuitry other than the sync section. As shown in Figure 9-4, one output from the sync amplifier proceeds into the horizontal AFC section. Sync lock is dependent on normal operation of the AFC circuitry, which in turn is dependent on normal operation of the horizontal oscillator section. Faulty horizontal sync lock can result, for example, from a defect in the comparison-pulse branch. The bottom line is that "tough-dog" sync troubleshooting problems may be located in a circuit section other than the sync section per se.

VERTICAL SYNC, OSCILLATOR, AND DEFLECTION SYSTEM

An example of a widely used vertical sync and deflection configuration is shown in Figure 9-5. Observe that the output from the sync clipper branches into the vertical integrator and into the horizontal AFC section. The vertical-hold control adjusts the free-running frequency of the vertical oscillator so that it will

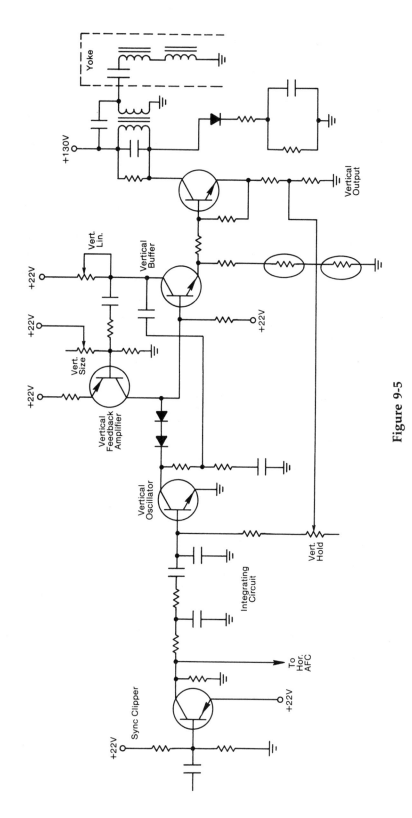

Figure 9-5

Figure 9-5 (Continued)

Note: This example of a discrete vertical-sweep section shows how the output from the sync clipper drives an RC integrating circuit to develop the vertical-sync pulse for triggering the vertical oscillator. The vertical oscillator is basically a free-running multivibrator with amplified feedback (for rapid retrace). In turn, the sawtooth output from the vertical oscillator drives the vertical buffer with its vertical output transformer and peaking circuit. Finally, the output from the vertical-output transformer is fed to the vertical-deflection coils in the yoke.

Reproduced by permission of Prentice Hall, from Complete TV Servicing Handbook, *by Walter H. Buchsbaum.*

Figure 9-5

Example of discrete vertical sync, oscillator, and deflection system.

lock on the vertical sync-pulse output from the integrator. Locking requires that the free-running frequency of the multivibrator be slightly less than 60 Hz.

The troubleshooter will not find a vertical-feedback amplifier in all receivers—this transistor is often omitted in black-and-white deflection circuitry. Observe that the vertical-size control adjusts the bias (and gain) and collector voltage on the oscillator transistors. The output from the oscillator is shaped by an RC "sawtooth" circuit which includes the vertical-linearity control. Note that there is normally considerable interaction between the size and linearity controls.

Next, a typical integrated vertical deflection circuit is shown in Figure 9-6. It performs basically the same functions as the arrangement in Figure 9-5. Note that the vertical blanking pulse is applied in the grid-cathode circuit of the picture tube, and serves to blank out retrace lines that could otherwise appear in the raster (particularly at higher settings of the brightness control).

NOISE MINIMIZATION

The amount of noise that gains entry into the deflection oscillator circuitry is related to the sync-channel bandwidth. This bandwidth is established by the RC low-pass and high-pass circuitry at the output of the sync separator. Noise pulses are generally quite narrow. In turn, a noise pulse has a broad frequency spectrum from very low frequencies to very high frequencies. Consequently, the amplitude of the noise pulses that pass through the sync channel

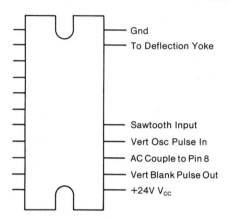

Gnd
To Deflection Yoke

Sawtooth Input
Vert Osc Pulse In
AC Couple to Pin 8
Vert Blank Pulse Out
+24V V_{cc}

Note: An integrated vertical deflection circuit is essentially the same arrangement as depicted in Figure 9-5, with active and passive devices and various resistors contained in the IC package. In the event of vertical malfunction, the troubleshooter checks the components that are external to the integrated circuit. Then, if the external components are ok, it is concluded that the fault is inside of the IC package. Observe that the example shown above is for a small-screen receiver. Integrated vertical deflection circuits for larger-screen receivers include a built-in heat sink which in turn is secured to a larger heat sink on the PC board.

Figure 9-6

Typical integrated vertical deflection circuit.

will be reduced if the bandwidth of the sync channel is reduced. There is, however, a practical limit to bandwidth reduction, inasmuch as the sync pulse will also be reduced in amplitude if the bandwidth of the channel is reduced beyond a certain point. As a rough guideline, the sync channel should pass approximately ten harmonics of the pulse repetition rate to avoid objectionable pulse attenuation.

Sync-channel circuitry with limited bandwidth is exemplified in Figure 9-7. Observe that transistor Q1 operates as a sync clipper in the common-collector mode, with signal-developed bias from R2 and C1. Resistor R1 functions to reduce the loading imposed on the picture detector by the sync separator. In turn, stripped sync pulses are fed to the base of Q2 which operates essentially as a sync amplifier in the common-emitter mode. Observe that R3, the load resistor for Q1, also serves as a bias resistor in combination with R4. Since Q2 is direct-coupled to Q1, its base resistance also operates in the bias network of Q1.

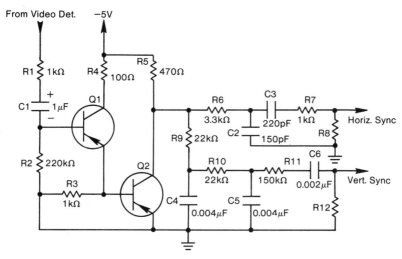

Note: This is an example of a widely used sync-channel arrangement using RC low-pass and high-pass filters with limited bandwidth for noise-pulse attenuation. The benefit of limited bandwidth is apparent during weak-signal reception whereby much more stable sync lock is realized than normally. Observe that if C_4 becomes open, the noise immunity of the vertical-sync channel will become poor. Similarly, if C_2 becomes open, the noise immunity of the horizontal-sync channel will become impaired.

Technical Note: Preliminary troubleshooting is usually accomplished to best advantage by means of comparison tests using a dual-trace scope.

Reproduced by permission of Prentice Hall, from TV Theory and Servicing *by Clyde Herrick.*

Figure 9-7

Example of sync-channel, low pass, and high-pass filters with limited bandwidth.

Observe that Q2 saturates on the peaks of pulse drive because it has a low collector voltage and a high value of load resistance (R5). Accordingly, the sync pulses that pass through Q1 and Q2 are limited in amplitude (have an unvarying output peak value).

Next, stripped sync pulses from Q2 are fed to a horizontal sync channel and to a vertical sync channel. Note that this horizontal channel is essentially a differentiating network consisting of C3, R7, and R8. However, the differentiating network is also preceded by the integrating circuit comprising R6 and C2. This integrating circuit has a time-constant of approximately 0.5 μs, with the result that the high-frequency response of the channel extends to approximately 150 kHz. This bandwidth permits horizontal sync pulses to pass with only slight attenuation, whereas most noise pulses are considerably attenuated. (Very narrow noise pulses are not attenuated, and are controlled by means of the AFC circuitry, as detailed below.)

Observe that comparable circuit design is employed in the vertical-sync channel (Figure 9-7) to provide limited bandwidth. A two-section integrator is provided, consisting of R9C4 and R10C5. The time-constant of the integrator is a compromise between minimum vertical-sync attenuation and maximum noise-pulse attenuation. Note that R11 serves as an isolating resistor between the integrator and its following load circuit. This load circuit starts with a differentiating circuit comprising C6 and R12. This differentiating circuit has a time-constant of approximately 300 μs, and thereby limits the low-frequency response of the channel to approximately 50 Hz. In other words, any low-frequency noise or transient voltages with a repetition rate less than 50 Hz are rejected.

POINTERS ON CIRCUIT ACTION

Ideally, the troubleshooter would be able to completely describe the circuit action in each subsection of any network that he or she encounters. Practically, the ability of even the most experienced troubleshooter is considerably less than ideal. Nevertheless, even the beginning troubleshooter has a general idea of the circuit action in simple networks—particularly DC networks. As the beginner gains practical experience, he or she builds up increasing knowledge of circuit action in AC networks. A comprehensive description of circuit action in any configuration includes notes concerning the changes in circuit action that result from off-value components and specified device faults.

Consider the three basic transistor configurations shown in Figure 9-8. These are the common-emitter, common-base, and common-collector configurations. With reference to Figure 9-7, we observe that Q1 operates in a CC configuration, and Q2 operates in a CE configuration. It is seen in Chart 9-1 that the CE and CB configurations may have a voltage gain up to 1000 times, or more, depending on the value of the load resistance. However, the CC configuration has a gain that approaches 1, but never exceeds unity, regardless of the load-resistance value.

However, in deflection systems, we are concerned with power levels and power gains, instead of voltage gains. The power gain of a transistor is equal to its voltage gain multiplied by its current gain. In Chart 9-1 the CE and CC configurations have a current gain in the order of magnitude of the transistor's

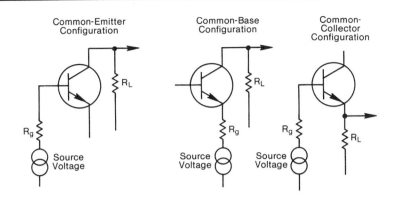

Common-Emitter Configuration   Common-Base Configuration   Common-Collector Configuration

Note: In the common-emitter configuration, the base is driven and the output is taken from the collector. In the common-base configuration, the emitter is driven and the output is taken from the collector. In the common-collector configuration, the base is driven and the output is taken from the emitter.

Technical Note: Occasionally, we encounter a variation of the common-emitter configuration in which an emitter resistor is included with two outputs, one taken from the collector and another taken from the emitter. This arrangement is sometimes called a paraphase inverter; the collector and emitter outputs are 180° out of phase (single-ended drive is used and double-ended output is obtained).

Figure 9-8

Basic common-emitter, common-base, and common-collector configurations.

beta value, whereas the CB configuration has a gain that approaches 1 but cannot exceed unity. Observe that, in turn, the power gain is highest for the CE configuration.

Circuit action can depend greatly on the input resistance and the output resistance of a circuit in various networks where loading is involved. With reference to Chart 9-1, it is seen that the common-collector configuration has the highest input resistance. However, the common-base configuration has the highest output resistance (at least for higher values of generator resistance). The graphs in Chart 9-1 show how the voltage gain, current gain, and power gain of a stage vary with off-value components. The graphs also show how the input and output resistances of the stage vary with off-value components.

BIASED NEON-BULB QUICK CHECKER

A useful quick checker for comparatively weak fields is shown in Figure 9-9. When a neon bulb is biased just below its extinction point, it becomes a much more sensitive field indicator than an unbiased bulb. Thus, the quick checker can be used to determine whether deflection current is flowing in the vertical-deflection coils of the yoke. If normal deflection current is flowing, the

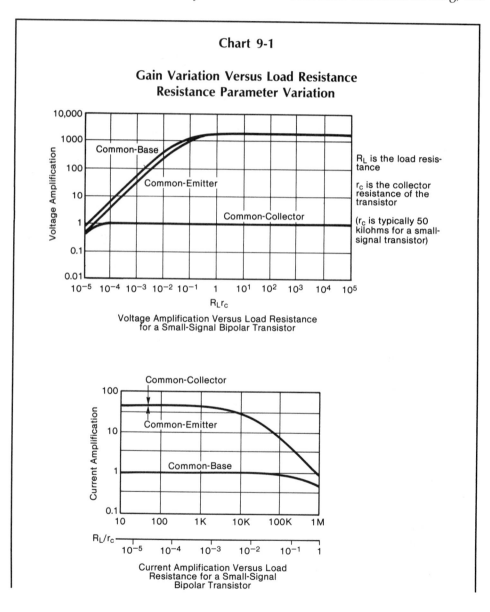

Chart 9-1

Gain Variation Versus Load Resistance
Resistance Parameter Variation

R_L is the load resistance

r_c is the collector resistance of the transistor

(r_c is typically 50 kilohms for a small-signal transistor)

Voltage Amplification Versus Load Resistance for a Small-Signal Bipolar Transistor

Current Amplification Versus Load Resistance for a Small-Signal Bipolar Transistor

Chart 9-1 (continued)

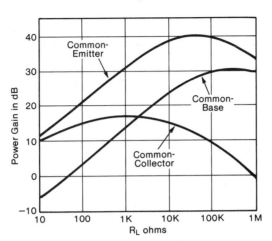

Power Gain Versus Load Resistance
for a Small-Signal Bipolar Transistor

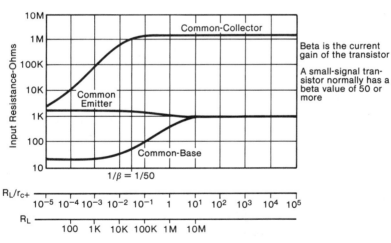

Input Resistance Versus Load Resistance for a
Small-Signal Bipolar Transistor

Chart 9-1 (continued)

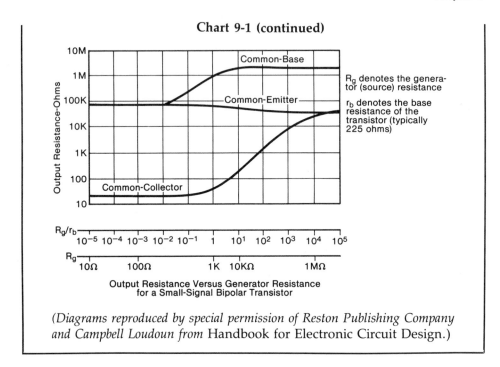

R$_g$ denotes the genera-
tor (source) resistance

r$_b$ denotes the base
resistance of the
transistor (typically
225 ohms)

**Output Resistance Versus Generator Resistance
for a Small-Signal Bipolar Transistor**

*(Diagrams reproduced by special permission of Reston Publishing Company
and Campbell Loudoun from* Handbook for Electronic Circuit Design.)

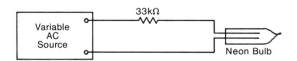

Note: This handy quick checker consists of a variable AC source such
as a variable power transformer, a 33kΩ resistor, and a neon bulb
such as an NE-15. The output voltage from the AC source is increased
until the neon bulb glows. Then, the output voltage is reduced until
the neon bulb barely extinguishes. In turn, the bulb is in its most
sensitive indicating state; the bulb will glow even if it is in the vicinity
of a comparatively weak field. Accordingly, if the neon bulb is moved
over and around the PC board or the picture tube in a TV receiver, the
bulb will glow in various regions. By way of comparison, if the neon
bulb is unbiased, it will glow only in a strong field, such as in the
proximity of the high-voltage transformer.

Figure 9-9

Arrangement of the biased neon quick checker.

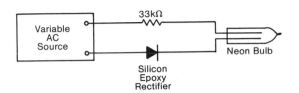

Note: A rectifier connected in series with the neon bulb makes the arrangement considerably more sensitive to fields in its vicinity. Observe that when a rectifier is used, only one electrode in the bulb will glow. Note also that when rectified voltage is used, the orientation of the bulb in a field may have more or less effect on the indication. This is the result of differing positive and negative peak voltages in pulse fields. In other words, the neon bulb may need to be suitably oriented in a pulse field so that a correct assisting polarity is present between electrodes.

Figure 9-10

**A rectifier can be used in series with the neon
bulb to increase its field sensitivity.**

biased neon bulb glows when brought near the coils. On the other hand, if little or no deflection current is flowing, the bulb remains dark.

When the extinction voltage level is precisely set, you may observe that only one electrode remains glowing in the bulb. In this state, the neon bulb is considerably more sensitive than before, and even a relatively weak field will cause both electrodes to glow. Note that if the neon bulb cannot be readily adjusted to leave only one electrode glowing, you can use a rectifier in series with the bulb, as shown in Figure 9-10. As the applied voltage is increased, one electrode will start to glow. Then, if the applied voltage is reduced to just below the extinction level, the bulb will be very sensitive to weak fields. It may be found desirable to reduce the applied voltage a bit more, so that the bulb becomes somewhat less sensitive.

Note that if the neon bulb is operated in a highly sensitive state, it will "strike" at a considerable distance away from the source of the field. The location of the source cannot be "pinpointed" unless the sensitivity of the neon bulb is reduced to a point that the bulb does not "strike" until it is brought close to the source. Observe that the general location of the source can be determined, even though the bulb is glowing at a distance, by watching the intensity of the glow as the bulb is moved about. As the bulb is brought nearer the source, its brightness increases.

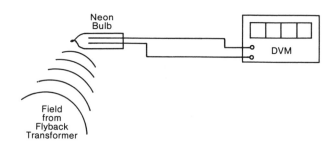

Note: This is an instructive experiment that demonstrates the value and the polarity of DC voltage generated by a neon bulb when placed in various regions and various orientations with respect to the field of a flyback transformer. Observe that no voltage is generated unless there is a glow in the neon bulb. The polarity of the generated voltage depends on the orientation of the bulb in the field. The value of the generated voltage depends on both the region of the field in which the bulb is placed, as well as the orientation of the bulb. In a typical situation, the generated voltage may range from zero up to 10 volts.

Figure 9-11

Experiment demonstrates the DC voltage generated by a neon bulb placed in a field.

EXPERIMENT WITH DC VOLTAGE GENERATED BY A NEON BULB

A helpful experiment with an ionized gas tube is shown in Figure 9-11. It shows the polarity and the value of the voltage that is generated by a glowing neon bulb as it is moved about in the field of a flyback and high-voltage transformer. When the neon gas glows, it is undergoing an ionized state in which atoms are stripped of electrons due to the field force. The lost electrons have a negative charge, and the ionized neon atoms have a positive charge. The experiment shows that a potential difference up to 10 volts can be generated when the neon bulb is placed in a suitable region of the field, with suitable orientation.

10

FOLLOW-UP TELEVISION TROUBLESHOOTING METHODS

Quick Check for Horizontal Sync Pulse • Point-to-Point Impedance Checker • Integrated Circuit Pointers • Horizontal and Vertical Deflection Integrated Circuit • Horizontal AFC Stage • Field Intensity Checker • Tuned-Coil and DVM Field-Intensity Checker • Comparison Temperature Checking • Progressive Trouble Symptom Development

QUICK CHECK FOR HORIZONTAL SYNC PULSE

When the picture will not lock in horizontal sync, and the horizontal-hold control can be adjusted to momentarily "frame" the picture, you can conclude that the fault is not in the horizontal oscillator section, but rather in either the automatic frequency control (AFC) section, or in the horizontal sync circuit. With reference to Figure 10-1, it is seen that if the AFC section is defective, the horizontal sync pulse is present but is not controlling the oscillator frequency. On the other hand, if the horizontal sync circuit is defective the horizontal sync pulse will be stopped and will not be present in the AFC/oscillator section.

The troubleshooter can often save time in localizing the trouble area by means of a quick check for the presence (or absence) of the horizontal sync pulse in the AFC/oscillator network. This is easily done as depicted in Figure 10-1. A mini-amp/speaker is connected at the output of the AFC section via a 50-kilohm resistor. If a tone is heard from the mini-amp/speaker, the troubleshooter concludes that the trouble will be found in the AFC section. On the other hand, if

Figure 10-1

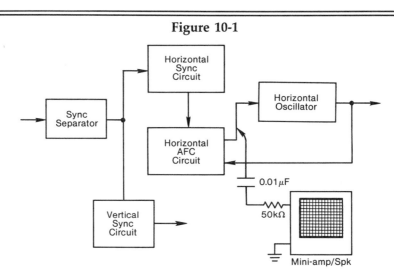

Note: This is a quick check to determine whether the horizontal sync pulse is being stopped in the horizontal sync circuit. (The horizontal oscillator is operating off-frequency and cannot be locked.) Observe that an incoming sync pulse (if present) will beat with the oscillator

waveform and will produce an audio tone due to the nonlinear circuit characteristics. However, if the horizontal sync pulse is not entering the AFC section, only a 15,750-Hz frequency is applied to the mini-amp/spk and an audio tone is not produced.

Observe that a blocking capacitor is used to prevent drain-off of DC voltage from the AFC circuit, and a 50-kilohm series resistor is included to minimize loading of the AFC circuit.

Note also, that if the picture did not lock in vertical sync, the troubleshooter would conclude that the fault is in the sync-separator section.

Figure 10-1

Horizontal sync pulse beats with off-frequency oscillator waveform to produce an audible tone.

the mini-amp/speaker does not output an audio tone, the troubleshooter concludes that the trouble will be found in the horizontal sync circuit.

Observe in Figure 10-1 that the quick check could "miss" a fault in the horizontal AFC circuit that presented an open-circuit or a short-circuit to the incoming sync pulse from the horizontal sync circuit. Accordingly, a somewhat more elaborate quick checker as that shown in Figure 10-2 may be used to definitely indicate the presence or absence of output from the horizontal sync circuit, provided only that the input terminal of the horizontal AFC circuit is not short-circuited to ground (a comparatively remote possibility). See also Figure 10-3.

Figure 10-2

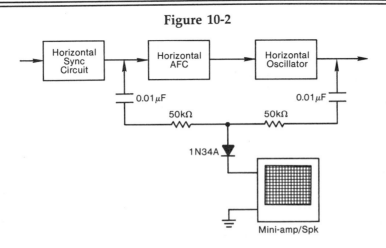

Figure 10-2 (Continued)

Note: This sync quick-checker arrangement mixes a sample of the output from the horizontal sync circuit with a sample of the output from the horizontal oscillator circuit. (The horizontal oscillator is operating off-frequency and cannot be locked.) Observe that if there is a horizontal sync-pulse output from the horizontal sync circuit, an audio beat note will be generated via the 1N34A diode and will be reproduced by the mini-amp/spk. In turn, the troubleshooter concludes that the fault is in the AFC section. On the other hand, if the mini-amp/spk is silent, the troubleshooter concludes that the fault is in the horizontal sync circuit.

Figure 10-2

Elaborated sync quick-checker arrangement.

Figure 10-3

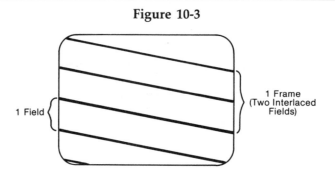

Note: When the horizontal oscillator is off-frequency, the number of dark diagonal bars in the image indicates how great the frequency error is, and the slope of the diagonal bars shows whether the oscillator is running too fast, or too slow. As an illustration, there are four diagonal bars in the diagram above. Therefore, the frequency error is equal to 4 × 60, or 240 Hz. Since the diagonal bars slope downhill to the right, the oscillator is running too fast (15,750 + 240, or 15,790 Hz). If the bars sloped uphill to the right, the oscillator would be running too slow (15,750 - 240, or 15,510 Hz).

It may also be observed that each field is followed by a vertical-sync pulse (diagonal bar). Successive fields are interlaced to provide maximum picture detail. A frame comprises two successive fields. Thus, a field is one-half of a TV image. Each field consists of $262\frac{1}{2}$ lines and is scanned in 1/60 second. In turn, 30 frames are formed each second.

Figure 10-3 (Continued)

As detailed subsequently, when a TV picture is "frozen," a chosen frame is scanned over and over. The frame comprises 525 interlaced lines.

Figure 10-3

Number of diagonal bars on the screen indicates the horizontal-oscillator frequency.

INTEGRATED CIRCUIT POINTERS

Troubleshooters may encounter integrated circuits in sync and deflection circuitry, and it is helpful to keep the basic functions of these devices in mind. For example, an IC signal processor is depicted in Figure 10-4. It performs AGC and sync-separation functions, including noise-reduction circuitry. A more elaborate example of integrated circuits in this general category is shown in Figure 10-5. In addition to the features noted above, this IC includes AFC, horizontal-oscillator, and horizontal-deflection preamp devices with various resistors. Observe that the circuitry employed by these integrated circuits is much the same as in discrete arrangements.

Figure 10-4

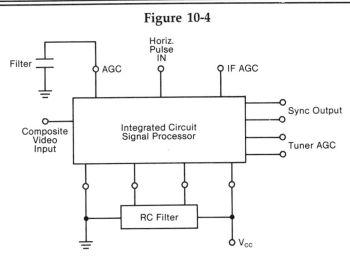

Note: This IC signal processor is used as a sync separator and noise-reduction arrangement in TV receivers. It is driven by a composite video signal from the video amplifier and by a keyer pulse from the horizontal-output circuit. The signal processor separates the sync

Figure 10-4 (Continued)

pulses from the composite video signal, reduces noise-pulse interference, and develops AGC control voltages for the tuner and the IF
section. Sync-pulse outputs are provided in either positive or negative polarity. Tuner AGC outputs are also provided in either positive
or negative polarity.

Figure 10-4

Features of integrated-circuit signal processor.

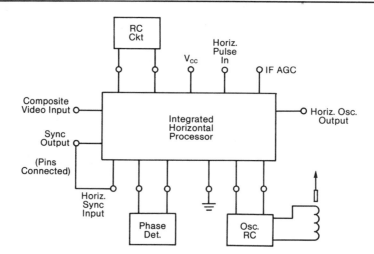

Note: This integrated circuit is used as a sync separator, AGC source,
horizontal AFC, oscillator, and preamp for the horizontal-deflection
section in TV receivers. It also includes noise-reduction circuitry. The
section marked RC Ckt contains the noise-reduction RC configuration. An external AGC filter arrangement is required with this IC. An
external phase-detector network is used for the horizontal-AFC function; it includes resistors, capacitors, and two diodes. The horizontal
oscillator employs an external RC circuit, and its frequency is determined by a tuned coil.

Figure 10-5

Features of an IC horizontal processor.

HORIZONTAL AND VERTICAL DEFLECTION INTEGRATED CIRCUIT

Troubleshooters will sometimes encounter the horizontal and vertical deflection IC depicted in Figure 10-6. It provides horizontal and vertical sweep waveforms to drive output stages. An unusual feature of this arrangement is that it does not employ horizontal and vertical hold controls. The horizontal oscillator is a 503.5-kHz ceramic resonator; it is followed by counter circuitry and its frequency is counted down in increments of 32 to produce the 15,750-Hz horizontal driving waveform. A timing circuit inside the IC also provides automatic lock of the 15,750-Hz and 60-Hz driving waveforms.

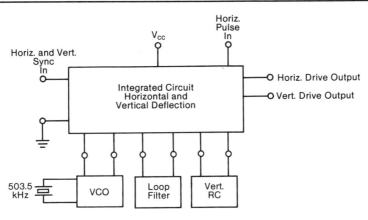

Note: This integrated circuit is used to provide horizontal and vertical deflection drive waveforms. It features a 503.5 kHz ceramic resonator operating in a voltage-controlled oscillator circuit. The horizontal flyback pulse input is compared with the horizontal sync input in a conventional phase detector to develop the control voltage for the VCO. An external loop filter smooths the control voltage to practically pure DC. The 503.5-kHz oscillator output is counted down in increments of 32 to provide the 15,750-Hz horizontal output. A timing circuit processes twice the horizontal output frequency and the incoming vertical sync pulse, and the vertical RC network serves to pass the vertical pulse but to reject the horizontal pulses. Due to the timing circuit, the 60-Hz vertical output is automatically and precisely locked in with the horizontal output, thus ensuring accurate interlace in the raster.

Figure 10-6

Features of a horizontal and vertical deflection configuration in integrated-circuit form.

Due to the complexity of this type of circuitry, troubleshooting is done to best advantage on a comparison basis with a similar receiver that is in normal operating condition. Voltage and resistance measurements at the IC pins will usually pinpoint the faulty circuit, and checks of the associated external components will then show whether the IC is at fault, or whether one of the external components is defective.

HORIZONTAL AFC STAGE

It is helpful to recognize the widely used horizontal AFC arrangements. The transistor type is exemplified in Figure 10-7; it operates as a phase detector and compares the phase of the incoming sync pulses with the phase of a comparison sawtooth from the horizontal-output section. Another AFC arrangement is actually quite similar, except that it employs a pair of diodes, instead of using the emitter and collector junctions in a transistor.

Figure 10-7

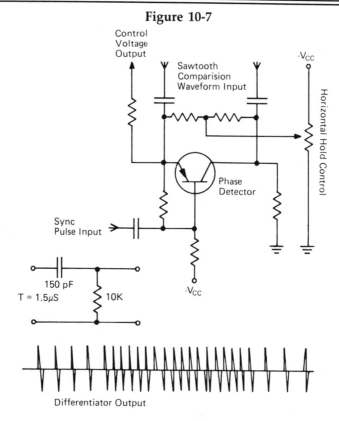

Differentiator Output

Note: The horizontal phase detector develops a DC control voltage which adds to or subtracts from the fixed bias on the horizontal-oscillator transistor, and thereby corrects any tendency of the hori-

Figure 10-7 (Continued)

zontal oscillator to drift off frequency. This control voltage may be either positive or negative.

Negative-going differentiated sync pulses applied at the base of the phase detector transistor bring the transistor out of cut-off into conduction. The sync pulse input capacitor is charged at this time by the flow of base current, and holds the transistor in cut-off during the interval between sync pulses.

As noted in the diagram, the collector-to-emitter voltage is a sawtooth waveform obtained from the horizontal-output stage. This collector-to-emitter voltage is negative during the first half of the retrace interval, and the transistor conducts in the usual mode. However, during the latter half of the retrace interval, the collector goes positive and now functions as an emitter. If the horizontal oscillator is phased properly, the center of the sync pulse precisely straddles the point where the sweep signal passes through zero during the retrace interval.

Under this condition, during the time that the sync pulse holds the transistor in conduction, current flows first from emitter to collector; then the current reverses and flows from collector to emitter. Since the peak current amplitude in this situation is the same for both directions of flow, the resulting control voltage output is zero.

However, if the horizontal oscillator starts to pull, the sync pulse will arrive a bit early, or a bit late with respect to the point where the sweep signal passes through zero during the retrace interval. In turn, the current flow through the phase detector is greater in one direction and less in the other direction. Consequently, a positive or a negative control-voltage output is produced. This control voltage functions to bring the horizontal oscillator back on frequency.

Figure 10-7

A widely used horizontal-AFC arrangement.

A widely used arrangement that uses a pair of diodes instead of a transistor for development of the oscillator control voltage is shown in Figure 10-8. Observe that a flyback comparison pulse is applied to the diodes with their associated RC circuitry. The result is production of a semi-sawtooth waveform that is "in step" with the horizontal deflection voltage. Observe also that a horizontal sync pulse from the sync amplifier is injected at the juncture of the

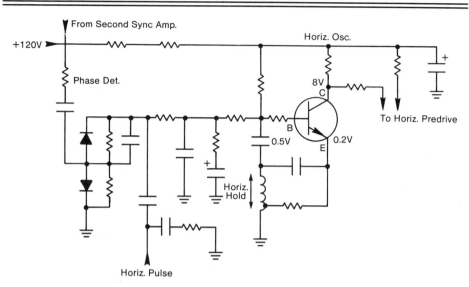

Horiz. Pulse

Note: This horizontal oscillator/AFC arrangement is used in many TV receivers. Observe that the average base-emitter bias voltage is 0.3 volt. If the oscillator tends to drift off-frequency, the diodes respond with a DC control voltage that changes the base-emitter bias voltage as required to bring the oscillator back on-frequency. In case of malfunction, electrolytic capacitors are ready suspects. Note that the diodes must be well matched, or the sync lock will not be "tight."

Figure 10-8

A widely used AFC arrangement with a pair of control diodes.

two diodes. In turn, the AFC network processes a sawtooth waveform with a superimposed pulse that "rides" higher or lower on the sawtooth, depending on the relative timing of the pulse and the sawtooth.

Note that when the sync pulse moves up the sawtooth wave, the peak voltage across one diode increases, and the peak voltage across the other diode decreases. Accordingly, one diode now conducts more than the other diode, and the base bias voltage on the oscillator transistor also changes. If the base voltage goes more positive, the oscillatory frequency decreases, and vice versa. Thus, the phase relations of the horizontal sync pulse and the sawtooth wave operate to pull the oscillator back on-frequency whenever it tends to run too fast or too slow. In preliminary fault analysis, the troubleshooter should observe whether the oscillator is running too fast or too slow (refer to Figure 10-3).

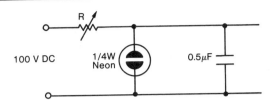

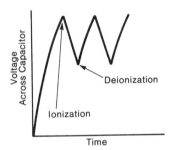

Note: This field-intensity quick checker functions as a variable frequency oscillator with a neon bulb such as the NE-15 as an indicator. The value of R may be in the order of 1 megohm, and should not be less than approximately 30 kilohms to avoid damage to the neon bulb. The circuit functions as a semi-sawtooth oscillator, and the neon bulb flashes on each ionization peak. Its flashing rate increases when the bulb is brought into the field of a flyback and high-voltage transformer, or into the field of the vertical-deflection coils.

Figure 10-9

Arrangement of a neon-bulb oscillator for making quick checks of field intensity.

FIELD INTENSITY CHECKER

A neon bulb is a handy high-voltage section checker. If the neon bulb is brought into the vicinity of the flyback and high-voltage transformer, the bulb will glow if the transformer is operating normally. The field intensity can be roughly checked by observing the brightness of the glowing bulb. However, this is only a qualitative estimation and does not provide precise indication of the field intensity. In preliminary troubleshooting procedures, and particularly in comparison tests, it is quite helpful to have a precise indication of field intensity.

This is easily accomplished with the use of a neon-bulb oscillator arrangement, as shown in Figure 10-9.

This checker operates on the basis of oscillating frequency. For example, if the neon bulb flashes off and on at the rate of one pulse per second in the absence of a field, its flashing rate will increase to 10 pulses per second when the bulb is placed in a weak field. As the field strength increases, the flashing rate of the bulb also increases, and soon the troubleshooter's persistence of vision makes it appear that the bulb is glowing steadily. For this reason, the chief usefulness of the neon-bulb oscillator is in making comparison checks in relatively weak fields.

TUNED-COIL AND DVM FIELD-INTENSITY CHECKER

Most precise measurements of relative field intensity can be made by means of the tuned-coil and DVM arrangement shown in Figure 10-10. It operates at 15,750 Hz, and has high sensitivity. The coil may be a replacement type for a TV receiver that uses a sine-wave horizontal oscillator, or it may be scavenged from a discarded TV receiver. This type of quick checker is helpful in preliminary troubleshooting procedures because it shows whether a dark-screen trouble symptom is being caused by a fault in the AC section of the flyback and high-voltage circuitry, or by a fault in the DC section.

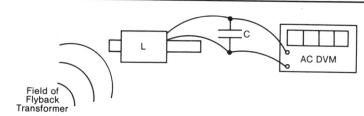

Note: This checker provides the most precise field comparison data in preliminary troubleshooting of the horizontal deflection system. It utilizes a coil L tuned to 15,750 Hz by capacitor C. The coil may be salvaged from a TV receiver that employs a sine-wave horizontal oscillator. (When the coil resonates at 15,750 Hz, maximum indication is obtained on the DVM.) Observe that the output voltage from the coil is greatly dependent on the orientation of the coil in the field. For example, when the coil is placed near the flyback transformer and suitably oriented, the DVM may read 300 V. On the other hand, when the coil is rotated 90°, the DVM may read practically zero. In comparison tests, it is essential to place the coil in precisely the same position in each test.

Figure 10-10

Tuned-coil and DVM field-intensity checker.

PROGRESSIVE TROUBLE SYMPTOM DEVELOPMENT

As mentioned previously, trouble symptoms can result from deteriorated filter capacitors. Another series of trouble symptoms is caused by power-supply defects that involve rectifier diodes or resistors. (See Figure 10-11.) In this situation, the power-supply output voltage is subnormal, although filtering is normal. The progressive trouble symptoms that develop depend to some extent on individual receiver design. However, the following symptoms are typical:

1. Picture and sound reproduction are normal at power-supply output voltages corresponding to an input voltage range from 117 down to 100 volts rms.

Figure 10-11

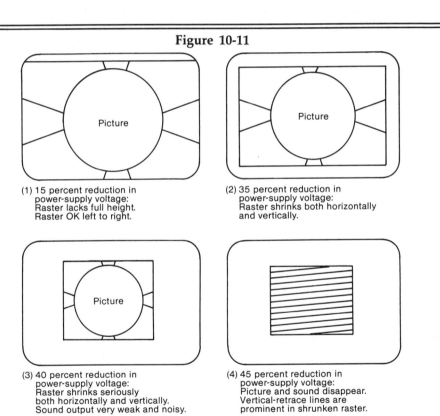

(1) 15 percent reduction in power-supply voltage: Raster lacks full height. Raster OK left to right.

(2) 35 percent reduction in power-supply voltage: Raster shrinks both horizontally and vertically.

(3) 40 percent reduction in power-supply voltage: Raster shrinks seriously both horizontally and vertically. Sound output very weak and noisy.

(4) 45 percent reduction in power-supply voltage: Picture and sound disappear. Vertical-retrace lines are prominent in shrunken raster.

Note: Both horizontal-deflection and vertical-deflection operation are progressively affected by subnormal power-supply voltage. Sync operation can be maintained by readjustment of the horizontal-hold and vertical-hold controls. In a typical receiver, both picture and sound reproduction stop abruptly when the power-supply voltage falls by 45 percent.

Figure 10-11 (Continued)

A "stone dead" receiver arouses an immediate suspicion of a catastrophic power-supply failure.

Figure 10-11

**Progressive trouble-symptom development
resulting from subnormal power-supply voltage.**

2. At 100 volts, the first trouble symptom that appears is a noticeable reduction in picture height at the top of the screen.

3. At 80 volts, additional trouble symptoms are observed; the reduction in picture height is substantial and is apparent at both the top and bottom of the screen. Also, there is a noticeable reduction in picture width at the left and at the right of the screen. The sound reproduction becomes weaker, although it is clear.

4. At 70 volts, the picture becomes seriously reduced in height and width. Horizontal and vertical sync lock become uncertain. Brightness and contrast are inadequate. The sound becomes quite weak and noisy. The sync, brightness, and contrast controls can be readjusted for satisfactory picture reproduction. However, the sound cannot be brought up to normal volume.

5. At 67 volts, the picture image suddenly disappears, although the raster remains visible. Sound reproduction and noise also stop. (The remaining raster is seriously reduced in height and width, and vertical-retrace lines are prominent.)

Consider next the progressive development of trouble symptoms in a basic amplifier arrangement. For example, with reference to Figure 10-12 it is shown how the DC collector voltage in a basic amplifier stage decreases substantially as a result of "collector leakage" (leakage from collector to base). In other words, the operating point of the transistor changes if collector-junction leakage occurs. Insofar as DC-voltage distribution is concerned, "collector leakage" upsets the base-bias voltage-divider relations. Even if the shift in operating point is not too serious, the stage gain with respect to the applied signal voltage will be reduced considerably. The reason for this reduction in gain is that collector-junction leakage functions as spurious negative feedback from collector to base with respect to the signal.

Another common cause of stage malfunction is drift in value of a resistor. As an illustration, the effect of resistance change in the collector load is shown in Figure 10-13 for a typical class-A stage. If the load resistance doubles in value, or if it decreases to half its normal value, the circuit operation is greatly

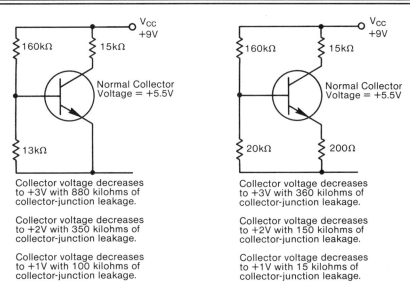

Collector voltage decreases to +3V with 880 kilohms of collector-junction leakage.

Collector voltage decreases to +2V with 350 kilohms of collector-junction leakage.

Collector voltage decreases to +1V with 100 kilohms of collector-junction leakage.

Collector voltage decreases to +3V with 360 kilohms of collector-junction leakage.

Collector voltage decreases to +2V with 150 kilohms of collector-junction leakage.

Collector voltage decreases to +1V with 15 kilohms of collector-junction leakage.

Note: These configurations are typical common-emitter amplifier arrangements with the widely used voltage-divider bias circuit. The first diagram employs no feedback; the second diagram employs emitter-resistor feedback. A common cause of subnormal stage gain is development of collector-base (collector-junction) leakage inside the transistor. This fault shows up prominently with respect to the collector DC voltage. Observe that the DC collector voltage decreases substantially as a result of collector-junction leakage. Note also that the amplifier is more tolerant of collector-junction leakage when emitter-resistor feedback is used.

Figure 10-12

Examples of collector-voltage shift due to collector-junction leakage.

affected. Observe that when the collector load resistance doubles, the stage gain also increases. However, distortion also occurs inasmuch as the stage is no longer operating in class A. Conversely, when the collector load resistance decreases, the stage gain also decreases.

Observe next the horizontal-oscillator configuration shown in Figure 10-14. This is technically an amplifier stage that supplies its own input by positive feedback via the coil between base, emitter, and ground. In normal operation, this type of circuit is usually biased into class B or into class C. The example is for class-C operation, with a base-emitter reverse bias of −1 volt. The

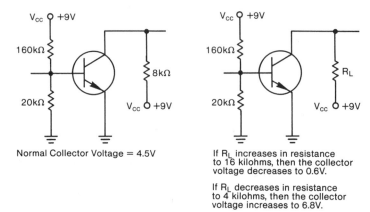

Normal Collector Voltage = 4.5V

If R$_L$ increases in resistance
to 16 kilohms, then the collector
voltage decreases to 0.6V.

If R$_L$ decreases in resistance
to 4 kilohms, then the collector
voltage increases to 6.8V.

Note: This is a typical class-A amplifier stage with a voltage-divider bias circuit. Optimum class-A bias occurs when the collector DC voltage is half way between V$_{cc}$ and ground. Observe that the collector voltage depends on the value of the load resistance, as well as the resistive proportions in the voltage divider. For example, if the load resistor increases in value, the collector voltage falls. On the other hand, if the load resistor decreases in value, the collector voltage rises. Note in the diagrams that if the load resistance doubles in value, the amplifier is no longer operating in class A; at 0.6 volt collector voltage, the stage is practically operating in class B.

Figure 10-13

Typical collector-voltage shift in class-A amplifier due to off-value load resistor.

Figure 10-14

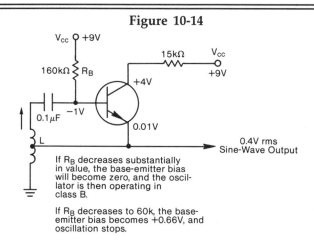

If R$_B$ decreases substantially
in value, the base-emitter bias
will become zero, and the oscillator is then operating in
class B.

If R$_B$ decreases to 60k, the base-
emitter bias becomes +0.66V, and
oscillation stops.

Figure 10-14 (Continued)

Note: This is a simple horizontal-oscillator arrangement that uses signal-developed self bias. In other words, the base conducts in pulses on the positive peaks of the sine-wave voltage across L. Base conduction results in charging the right-hand side of the 0.1 μF capacitor negatively. This negative charge is drained off between peaks by R_B, and the average DC bias on the base is −1 volt. This is a reverse bias voltage, or, the transistor operates in class C. Observe that if R_B decreases in value to 60 kilohms, the transistor then becomes forward-biased, and the sine-wave output drops to zero. Oscillation stops because the base draws excessive current (its input resistance becomes so low that L is damped excessively).

In case the transistor becomes heavily forward-biased, it will overheat and be damaged.

Figure 10-14

Basic horizontal-oscillator arrangement normally operates with reverse base-emitter bias.

−1 volt is an average value that represents the difference between the positive bias voltage applied through R_B, and the signal-developed self-bias voltage produced by base-current flow on positive peaks of the sinewave output voltage.

Note that if R_B drifts to a lower value, the oscillator will go to class-B operation (forward bias of 0.5 volt,) and if R_B drifts to a still lower value, such as 60 kilohms in this example, the oscillator will be forward biased and must operate in class A. However, under a forward-biased condition, the effective base-emitter input resistance becomes very low, and the oscillator is likely to be damped down excessively so that oscillation cannot occur. (This is the condition noted in the diagram.) The bottom line is that the base-emitter bias voltage is the most informative measurement in preliminary troubleshooting of a "dead" oscillator section.

11

COLOR CIRCUIT TROUBLESHOOTING

Progressive Trouble Symptom Development • General Considerations • Integrated Circuits • Chroma Signal Levels • Short-Wave Radio Receiver Responds to Chroma-Section Signals • Resistor Probe Improves SW Radio Receiver Tests • 3.58-MHz Quick Checker • Noise-Level CW Signal Tracing • Chroma Demodulation and Matrixing • Color Demodulator Troubleshooting • IC Functional Overview • Video Sweep Modulation

PROGRESSIVE TROUBLE SYMPTOM DEVELOPMENT

When troubleshooting color television receivers without service data, it is helpful to note trouble symptoms that require no prior knowledge of the circuitry that may be used in the receiver. Previous discussion of progressive trouble symptom development in black-and-white receivers under conditions of subnormal power-supply output voltage was provided in Chapter 10. As exemplified in Figure 11-1, the same general considerations apply also to color-TV receivers.

Figure 11-1

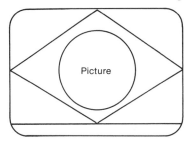

28 percent reduction in power-supply voltage:
Raster loses height (bottom of picture).

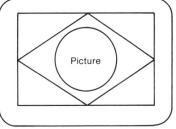

37 percent reduction in power-supply voltage:
Raster loses both height and width.
Picture and sound reproduction are still accaptable.

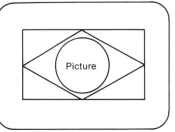

39 percent reduction in power-supply voltage:
Severe loss of raster height and width.
Picture dim. Sound is still acceptable.

43 percent reduction in power-supply voltage:
Screen becomes dark.
Sound is quite weak.

Figure 11-1 (Continued)

Note: In this example of subnormal power-supply output voltage in a widely used model of color-TV receiver, the first noticeable trouble symptom is vertical shrinking of the raster, followed by progressive vertical and horizontal shrinking. Then, the picture becomes dim, although color is still reproduced and the sound remains acceptable. Finally, the screen becomes dark and the sound becomes quite weak.

Figure 11-1

Example of progressive trouble symptom development in a color-TV receiver resulting from subnormal power-supply voltage.

If the raster lacks both height and width, the power-supply output voltage is likely to be found subnormal. However, if the raster has normal height but lacks normal width, the fault is most likely to be found in the horizontal-deflection section. Conversely, if the raster has normal width but lacks normal height, the fault is most likely to be found in the vertical-deflection section.

GENERAL CONSIDERATIONS

As shown in Figure 11-2, a color-TV receiver is more complex than a black-and-white receiver. In turn, if the trouble symptom is not simple (as in the case of subnormal power-supply voltage), it is usually advisable to try and locate a similar color-TV receiver in normal working condition to make comparison tests. However, if the troubleshooter has had previous experience with the same type of receiver, he or she may feel sufficient confidence to tackle a puzzling trouble symptom without the support of comparison testing.

As noted in Chapter 10, a "stone dead" receiver arouses the immediate suspicion of a catastrophic power-supply failure. This is usually a comparatively simple and straightforward troubleshooting job. It is good practice to start by checking receiver operation with a bench power supply for temporary substitution of the failed receiver power supply. This procedure can save otherwise wasted time if it is found, for example, that the receiver is still "stone dead" after power-supply voltage has been provided. In such a case, the troubleshooter may opt to stop work instead of investing further time on the problem.

In case the receiver is workable, although malfunctioning, the trouble symptom(s) should be analyzed critically. Sometimes the trouble area is apparent, as when the symptom is "black and white reception only—no color." In this situation, the troubleshooter turns his or her attention to the chroma circuitry. Again, the trouble area may be obscure, as when the symptom is "intermit-

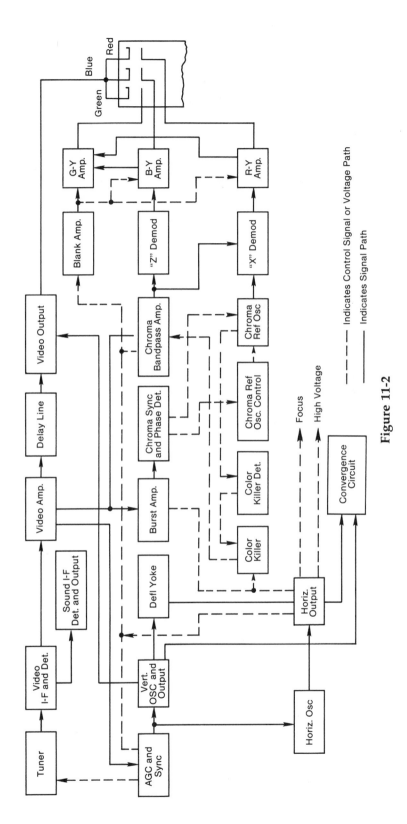

Figure 11-2

--- --- --- Indicates Control Signal or Voltage Path

——— Indicates Signal Path

231

Figure 11-2 (Continued)

Note: This is one of the standard color-TV receiver arrangements in which the black-and-white video (Y) signal is applied to the cathodes of the color picture tubes, and the R-Y, B-Y, and G-Y chroma signals are applied to the grids of the color picture tube. "X" and "Z" chroma demodulators are used; the G-Y signal is matrixed in the G-Y amplifier.

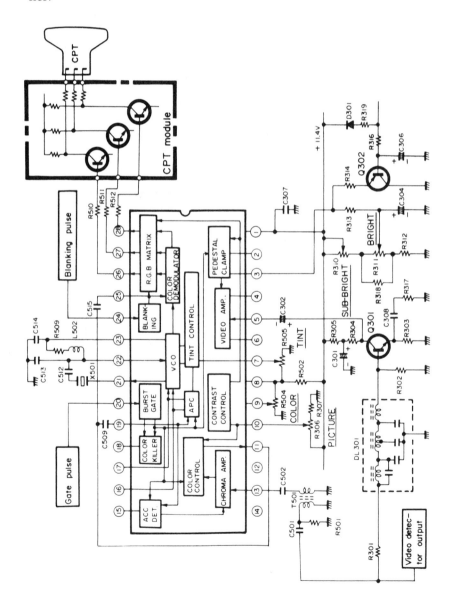

Figure 11-2 (Continued)

Note: This is another standard color-TV receiver arrangement in which the cathodes of the color picture tube are individually driven by R, G, and B signals (the grids are not driven). Observe that the black-and-white signal is formed from corresponding proportions of the R, G, and B signals which are sourced in an RGB matrix network. Troubleshooting this type of circuitry is accomplished to best advantage by means of comparison tests. However, if the troubleshooter is experienced in this area, fault localization is feasible by means of a voltmeter and various quick checkers, as explained subsequently.

Reproduced by permission of Prentice Hall, from Complete TV Servicing Handbook *by Walter H. Buchsbaum.*

Figure 11-2

Color-TV receiver arrangement: (a) receiver comprises all of the circuitry used in a black-and-white receiver, plus chroma signal processing circuits; (b) an arrangement in which the chroma section is an integrated circuit.

tent video." In either case, the troubleshooter is fortunate if the PC board is marked with functional section identifications.

INTEGRATED CIRCUITS

You will find integrated circuits in the chroma section, particularly in the more recent models of color-TV receivers. For example, a typical color-sync and color-killer IC is shown in Figure 11-3. Thus, if the trouble symptom is "black and white reception only—no color," the troubleshooter turns his or her attention to the chroma IC and its associated external components. The usual procedure is to check out the external components (especially the electrolytic capacitors). Then, if the external components "pass inspection," the IC will be replaced.

CHROMA SIGNAL LEVELS

The 3.58-MHz chroma signal is processed at a comparatively low level in modern color-TV receivers. For this reason, *it is not practical to use conventional peak-reading probes and voltmeters to trace the chroma signal.* Even a forward-biased peak-reading probe has limited usefulness in chroma-circuit troubleshooting. Instead, the technician generally uses chroma signal-injection tests to close in on a trouble area. The color picture tube is ordinarily used as an indicator. How-

Figure 11-3

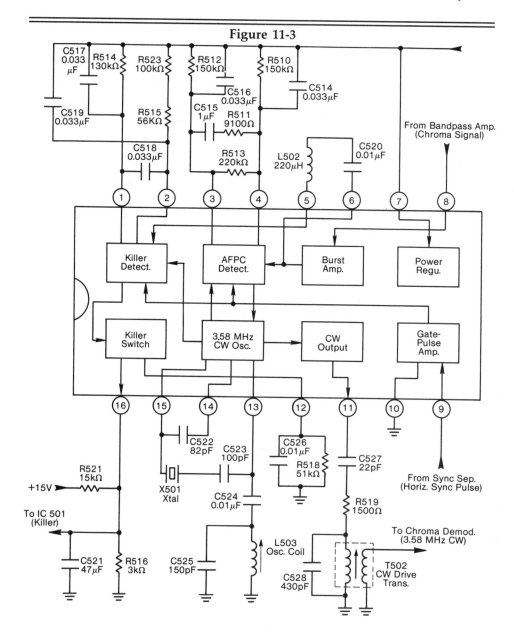

Technical Note: Preliminary troubleshooting in this type of circuitry is accomplished to best advantage by means of comparison tests with signal present, using a dual-trace scope. The troubleshooter does not need to know the detailed theory of circuit operation.

Figure 11-3 (Continued)

Reproduced by permission of Prentice Hall, from Complete TV Servicing
Handbook, *by Walter H. Buchsbaum.*

Figure 11-3

**A typical color-sync and color-killer integrated
circuit used in many color-TV receivers.**

ever, if the screen is dark, *the troubleshooter can use a peak-reading probe and a DVM
as an indicator at the picture tube cathode or at a grid.*

SHORT-WAVE RADIO RECEIVER RESPONDS TO CHROMA-SECTION SIGNALS

Chroma signal checking and signal tracing is accomplished to best advantage by means of an oscilloscope. However, if you do not have an oscilloscope, useful information can be obtained with a short-wave radio receiver
tuned to 3.58 MHz. For example, when the antenna rod is used as a probe and
brought into the vicinity of a 3.58-MHz circuit, a chroma signal will be picked up
and reproduced as a characteristic "roar." Thus, the presence or absence of
chroma signal can be determined in preliminary troubleshooting procedures.
Tests with a radio receiver are more informative if made on a comparative basis
with a similar receiver that is in normal operating condition.

Note that the chroma signal will be heard not only at 3.58 MHz but also
in the range from 3 to 4 MHz, and in some receivers from 2 to 4 MHz. The reason
for this "spread" of chroma signal frequencies is seen in Figure 11-4. Observe
that the bandpass amplifier in this example has a bandwidth from 3.1 to 4.1
MHz. In turn, sideband frequencies over this range can be picked up by the
radio receiver.

In addition, a high-performance color-TV receiver provides a bandpass
amplifier frequency range from 2.1 to 4.1 MHz. Accordingly, when this type of
color-TV receiver is being tested, the troubleshooter expects to find a frequency
"spread" from approximately 2 to 4 MHz. Insofar as the chroma reference
oscillator (color oscillator) is concerned, its output is inaudible on a short-wave
radio receiver because this is a 3.58-MHz continuous-wave signal. In other
words, only modulated chroma signals can be reproduced by an AM radio
receiver.

Chroma signals are video frequencies, and there is the possibility of
confusing chroma signals with black-and-white signals in simple quick checks.

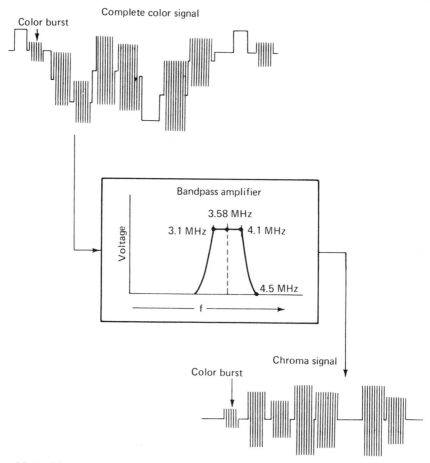

Note: These patterns illustrate the appearance of an NTSC color test signal before and after passage through the bandpass amplifier as displayed on an oscilloscope screen. This test signal contains the primary colors and their complementary colors. The chroma output signal from the bandpass amplifier consists of 3.58-MHz signal groups with various amplitudes and phases.

Reproduced by special permission of Reston Publishing Company, from Color Television Theory and Servicing *by Clyde Herrick.*

Figure 11-4

Bandpass amplifier typically passes chroma frequencies in the range from 3.1 to 4.1 MHz.

To avoid confusion, the troubleshooter should "size up" the circuitry, if necessary, to make certain that the end of the antenna rod is being brought into the bandpass network, for example, instead of the video-amplifier network.

More precise frequency information can be obtained if the troubleshooter uses a frequency counter. Three commonly available types of frequency counters are the 80-MHz, 250-MHz, and 520-MHz designs. A typical 80-MHz counter provides 20-mV sensitivity with a frequency range from 10 Hz to 80 MHz. A choice of gate time periods is available: 0.1 sec, 1 sec, or 10 sec. The input impedance is 1 megohm.

The 250-MHz and 520-MHz instruments have somewhat less sensitivity at frequencies above 100 MHz (typically 50 mV). They also provide a choice of high input impedance (1 megohm), or coax-cable impedance (50 ohms). Many frequency counters also provide period characteristics; the average frequency over a chosen time period is measured.

RESISTOR PROBE IMPROVES SW RADIO RECEIVER TEST

A wide range of chroma signal levels may be encountered under trouble conditions, and a signal level that is normally high may become low due to a component or device fault. As a result, a quick check for presence or absence of signal with a short-wave radio receiver could leave the troubleshooter in doubt concerning the status of chroma-circuit operation. In such a case, it is helpful to use a resistor probe with the rod antenna, as shown in Figure 11-5. The resistor increases the level of the desired signal into the antenna, and also avoids objectionable circuit loading.

Note that various short-wave AM radio receivers, such as the Radio Shack DX-55 provide an external-antenna jack, in addition to the rod antenna. This external-antenna jack can be used for connection of a test probe, although it provides no technical advantage over the simple arrangement shown in Figure 11-5. Although it might appear advantageous to use an isolating resistor and a coaxial cable as a test probe connected to the external-antenna jack, this is an impractical arrangement, due to the distributed capacitance of the coax cable. This is just another way of saying that when a 50-kilohm resistor feeds into the cable capacitance, a low-pass filter action results that practically "kills" a 3.58-MHz signal.

Similarly, it might appear advantageous to use a coax cable as a direct probe connected to the external-antenna jack. However, this is not a practical arrangement, due to the circuit loading imposed by the test cable. Stated otherwise, when a coax test lead is applied directly to an active circuit point in a chroma network, the distributed capacitance of the coax introduces shunt capacitance that disturbs circuit action both from bypassing action and from detuning action. The bottom line is that the test arrangement shown in Figure 11-5 is the most practical quick-check method.

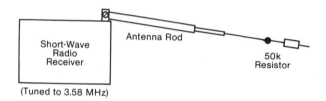

(Tuned to 3.58 MHz)

Note: When a short-wave radio receiver is used to check for presence or absence of signal in chroma circuits, the receiver responds to the signal when the end of the antenna rod is brought near the circuitry. If the signal level is comparatively low, it is helpful to increase the sensitivity of the test by using a 50-kilohm resistor as a probe at the end of the rod. When the resistor probe is touched to an active point in the circuit under test, the sound output from the radio increases substantially. Use of the resistor probe is additionally helpful in case there are strong stray fields from other sections of the TV receiver that are picked up by the antenna rod and interfere with the signal of interest. In other words, when the resistor probe is used, the effective signal-to-noise (signal-to-interference) ratio is improved significantly.

Figure 11-5

Resistor probe improves signal checks with short-wave radio receiver.

3.58-MHz CW QUICK CHECKER

In some trouble situations, most of the chroma system in a color receiver is "dead," although the color-subcarrier oscillator is operating and is producing a 3.58-MHz output. A short-wave radio receiver provides no information concerning 3.58-MHz oscillator operation, unless an additional signal input is supplied, as shown in Figure 11-6. The 3.58-MHz CW output from the signal generator will heterodyne with the output from the subcarrier oscillator, thereby producing a beat tone output from the radio receiver. The troubleshooter can quickly determine whether the color-subcarrier oscillator is functioning.

NOISE-LEVEL CW SIGNAL TRACING

If you do not have a signal generator available for heterodyne signal-tracing tests, noise-level quick checks can be made with a short-wave radio receiver, using the quieting effect of a CW signal as an indication of the presence or absence of signal. Thus, if the radio receiver is tuned to 3.58 MHz, and the

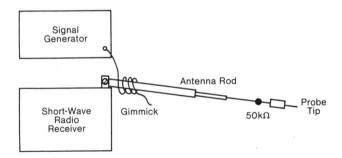

Note: This quick checker employs a short-wave AM radio receiver with the output from a signal generator coupled via a gimmick to the antenna rod of the receiver. The gimmick consists of several turns of insulated wire, and the antenna rod is terminated by a 50-kilohm isolating resistor. The signal generator is set for CW (unmodulated) output at 3.58 MHz, and the radio receiver is tuned to 3.58 MHz. Observe that if a 3.58-MHz CW signal is applied to the probe tip, the incoming signal will heterodyne with the generator signal and produce an audio tone from the radio receiver. (An audio tone is produced because it is extremely unlikely that the incoming signal and the generator signal will have exactly the same frequency.)

Localization of the color subcarrier oscillator is accomplished by switching the signal generator off and on. If the audio output from the radio receiver stops when the signal generator is switched off, the troubleshooter concludes that the probe tip is being applied in the subcarrier oscillator network.

Figure 11-6

Quick checker arrangement for tracing a 3.58-MHz continuous-wave signal.

end of the antenna rod is brought into the field of a 3.58-MHz CW oscillator, there will be no sound output from the radio. Observe, however, that the incoming CW signal has a quieting effect, and that the normal noise output from the radio decreases. This quieting effect indicates to the troubleshooter that there is an incoming CW signal at the frequency to which the radio receiver is tuned. As shown in Figure 11-7, it is often helpful to include an off-on push-button switch to compare the normal noise output from the radio receiver with the noise output at any time during quieting checks.

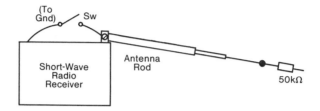

Note: A 3.58-MHz CW signal can be traced by means of its quieting effect, if a signal generator is not available for heterodyning. In other words, the radio receiver has more or less noise output in the absence of any signal input. Then, when the probe is applied to a 3.58-MHz CW signal source, AGC action reduces the IF gain in the receiver. Accordingly, the noise output decreases. This quieting effect can be used to trace a CW signal. As shown in the diagram, a push switch can be used with the radio receiver to momentarily short-circuit the antenna rod to ground in the receiver. In turn, the troubleshooter can compare the noise output at any time during a test with the basic noise level of the receiver by momentarily closing the switch.

If a communications-type short-wave receiver is available, it can provide more comprehensive quick checks than a simple receiver. For example, a communications receiver usually includes a built-in oscillator for heterodyning CW signals, and an audibility meter that shows comparative levels of incoming signals.

Figure 11-7

Push switch supplements quieting check for tracing 3.58-MHz CW signal.

CHROMA DEMODULATION AND MATRIXING

Troubleshooting of the chroma section is facilitated by a practical understanding of the chroma demodulation and matrixing processes. Observe that prior to transmission of the color-TV signal, the 3.58-MHz color subcarrier is suppressed (to minimize interference in the picture). Accordingly, the chroma demodulator circuitry reconstitutes the chroma signal by inserting the 3.58-MHz subcarrier into the incoming suppressed-carrier signal. The reconstituted signal is further processed by separation of its components in synchronous detectors. A synchronous detector, also called a product detector, is a combined phase and amplitude detector.

A color-difference signal, such as an R-Y, B-Y, or G-Y signal has a certain phase with respect to the burst signal; a color-difference signal also varies in amplitude. Note that chroma phase corresponds to the hue of a color, whereas chroma amplitude corresponds to the saturation of the color. R-Y and B-Y signals may be formed from demodulated X and Z color-difference signals, as detailed subsequently.

The plan of the basic R-Y/B-Y demodulator arrangement is shown in Figure 11-8. Observe that the chroma signal from the bandpass amplifier is applied to both of the demodulators; output from the subcarrier oscillator is applied to the B-Y demodulator in the B-Y phase, and is applied to the R-Y demodulator in the R-Y phase, thereby reconstituting the chroma signal with development of its R-Y and B-Y components.

This is termed *quadrature synchronous-detector action* since the R-Y and B-Y signal phases are separated by 90 degrees. Each demodulator conducts only for a brief interval at the peak of the injected 3.58-MHz voltage, as shown in Figure 11-9. In other words, synchronous detection is essentially a sampling of the

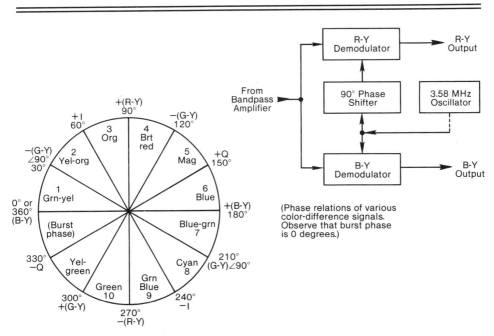

Reproduced by special permission of Reston Publishing Company, from Color-TV Theory and Servicing *by Clyde Herrick.*

Figure 11-8

Basic plan of an R-Y/B-Y demodulator section.

Figure 11-9

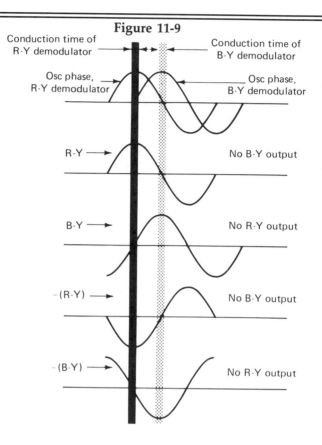

Quadrature synchronous-detector action.

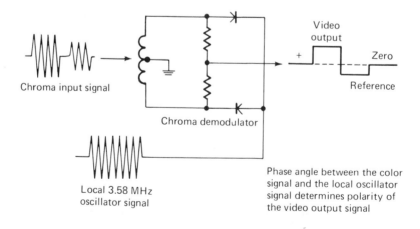

Figure 11-9 (Continued)

Reproduced by special permission of Reston Publishing Company, from Color-TV Theory and Servicing *by Clyde Herrick.*

Figure 11-9

Chroma demodulation signal-sampling process and basic synchronous demodulator circuit.

incoming signal at specified periods of time. Figure 11-9 shows that a synchronous detector is essentially a phase detector; it is also an amplitude detector, inasmuch as the output signal level "follows" the input chroma signal level.

As the R-Y signal component in Figure 11-9 is going through its peak value, the B-Y signal component is going through zero. Accordingly, the R-Y demodulator output consists of samples of the R-Y component only. Next, as the B-Y chroma-signal component is going through its peak value, the R-Y component is going through zero. Thus, the B-Y demodulator output consists of samples of the B-Y component only. Observe that the −(R-Y) and −(B-Y) signal components are sampled in a similar manner in Figure 11-9. As noted previously, a G-Y chroma-signal component is also developed—this may be a demodulator circuit action, with the color subcarrier signal injected in the G-Y phase, or, suitable proportions of the R-Y and B-Y demodulator outputs may be mixed in a G-Y matrix.

We will find various subsectional arrangements that provide the same end result (formation of the R, G, and B signals); for example, R-Y and G-Y signals may be demodulated, followed by a B-Y matrix. High-performance color receivers demodulate I and Q signals, from which R, G, and B signals are matrixed. In still another arrangement, R, G, and B signals are simultaneously demodulated and matrixed with the Y signal. Matrixing of X and Z signals to form R-Y, B-Y, and G-Y signals is shown in Figure 11-10.

Note that R is the matrix resistor in Figure 11-10. The X and Z signals are dropped in suitable proportion across R to form the G-Y signal; this is the same proportion required to form the R-Y and B-Y signals. The G-Y matrix first forms a −(G-Y) signal, which then passes through an inverter to produce the G-Y signal. As shown in Figure 11-11, the R, G, and B hues have specified phase angles, and they can be demodulated directly—this is done in receivers that apply R, G, and B signals directly to the picture-tube cathodes.

A helpful summary of the six basic arrangements for chroma demodulation and matrixing is shown in Figure 11-12. The troubleshooter may or may not have prior experience with these various types of chroma circuitry, and he or she may not have information concerning what form of demodulation and matrixing

Figure 11-10

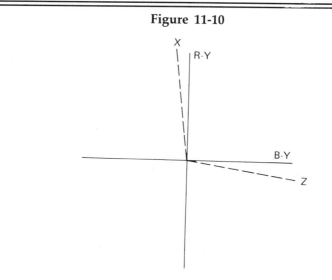

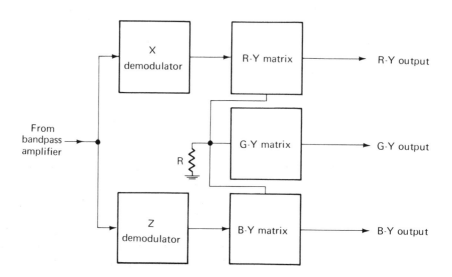

Note: It is seen from the phase diagram that if a small amount of the Z signal is mixed with the X signal, the result will be the R-Y signal. Similarly, if a small amount of the X signal is mixed with the Z signal, the result will be the B-Y signal.

Figure 11-10 (Continued)

Reproduced by special permission of Reston Publishing Company, from Color-TV Theory and Servicing *by Clyde Herrick.*

Figure 11-10

Matrixing arrangement for forming the R-Y, B-Y, and G-Y signals from demodulated X and Z signals.

is used in a particular receiver. For this reason, practical troubleshooting procedures may be limited to comparison tests with a similar receiver that is in normal operating condition.

IC FUNCTIONAL OVERVIEW

The functional arrangement for a widely used integrated circuit that processes the video-IF signal is shown in Figure 11-13. This IC performs the function of a three-stage broadband TV IF amplifier, detector, and video pre-amplifier. Automatic gain control (AGC) for the IF section and for the tuner is also provided in this IC. A horizontal keying pulse is used with a sample-and-hold circuit so that the AGC bias depends on the horizontal sync pulse amplitude. Thus, the AGC value is independent of camera-signal variations and noise from one sync-pulse interval to the next.

An independent adjustment is provided for setting the amount of delay for the AGC signal applied to the RF tuner. The AGC filter, consisting of resistors, capacitors, and a diode is a portion of the external network. One of the most important features of this IC is an internal voltage regulator that ensures constant power supply voltage to all stages. The four resonant circuits shown in Figure 11-14 are typical of video-IF bandpass networks, but they do not include the adjacent-channel and sound-IF traps. (In most TV receivers, these traps are a part of the filter connected between the tuner output and the IF input. Basic function parameters for the IC are:

1) *Nominal IF input signal.* Denotes the amplitude of the IF signal, over the flat portion of the frequency response curve, that will provide a nominal video output level between 1 and 5 V. A typical value is 400 mV.

2) *Distortion.* Denotes the amount of distortion at 50 kHz, with 80 percent amplitude modulation and a sync pulse amplitude equivalent to 30 rms mV. A typical value is 10 percent.

Figure 11-11

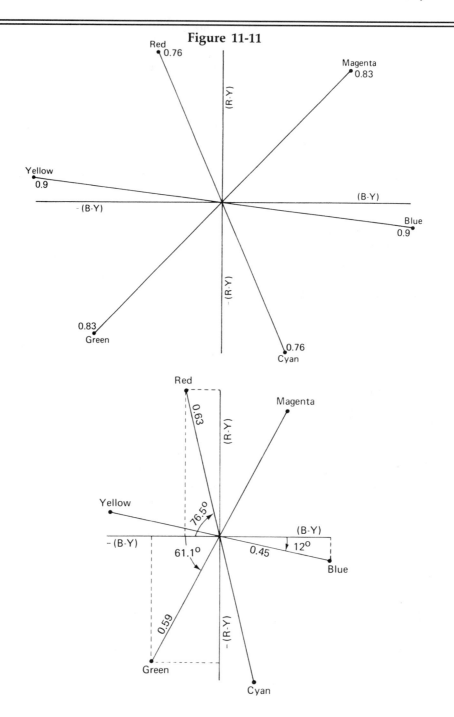

Figure 11-11 (Continued)

Note: The vectors shown above are termed unadjusted values for the fully saturated primary and complementary colors. They are the values used for color display by the color picture tube. On the other hand, the vectors shown below are called readjusted values. They are the values used in transmission and reception of the color signal in order to avoid overmodulation. The readjusted chroma values are outputted by the chroma demodulators. In turn, the readjusted chroma values are proportioned into unadjusted values by following resistive circuitry.

Reproduced by special permission of Reston Publishing Company, from Color-TV Theory and Servicing *by Clyde Herrick.*

Figure 11-11

Phase relations of the primary and complementary colors.

3) *Video output level.* Depending on the input signal, a range of 0.9 to 10 V output can be expected with 12 V_{cc}.

4) *Horizontal key pulse input.* Obtained through an external 100 kilohm resistor; between 25 and 35 peak-to-peak volts are required.

5) *Maximum power dissipation.* Up to 55°C, 750 mW may be dissipated.

Next, consider the functional arrangement of the luminance signal-processing integrated circuit shown in Figure 11-14. This IC provides the functions of equalizing the high- and low-frequency components of the video signal, clamping the signal to the proper black level, and mixing the vertical and horizontal blanking signals with the video signal. A power-transistor video-amplifier driver is usually required to drive the picture tube.

Because of the longer time required for processing the color signal, a tapped delay line of approximately 750 ns is used in modern color TV receivers. Three inputs from this delay line are applied to the luminance processor. A peaking amplifier and main video amplifier on this IC provide for the contrast control and peaking of the high video frequencies. A special black-level clamping circuit requires a horizontal pulse input at the clamp-inhibit terminal to prevent the video signal from being clamped to the sync level, which is more negative than the black level. The brightness control sets the level of the video output amplifier, which is also gated from the vertical and horizontal blanking pulses. Basic functional parameters for this IC are:

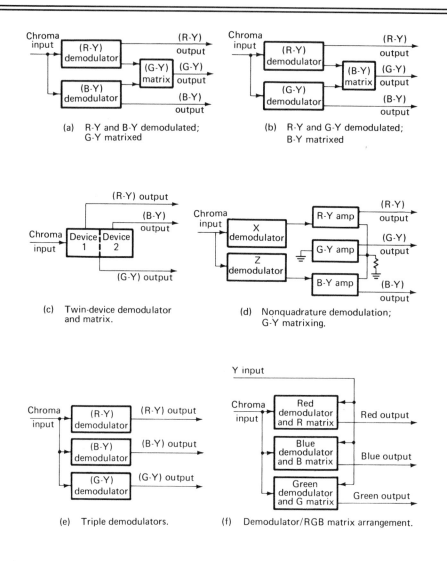

Figure 11-12

Six basic arrangements for chroma demodulation and matrixing, used in various color-TV receivers.

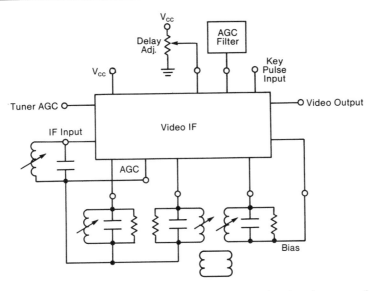

Technical Note: Although this is a comparatively simple example of integrated circuitry, it is a sufficiently complex system that its detailed action cannot be visualized by the practical troubleshooter. (It would be a waste of time to attempt to follow the circuit action in detail.) Accordingly, after a trouble symptom has been localized to the video-IF section, preliminary troubleshooting procedure generally consists of comparative DC voltage and resistance measurements at the IC pins, without regard to circuit actions that are involved. In most situations, these measurements will serve to show whether the IC is defective, or whether the fault is in an external component.

Figure 11-13

Functional arrangement of a widely used integrated circuit for processing the video-IF signal.

1) *Maximum power dissipation.* Rated at 750 mW at 55°C.

2) *Wideband gain.* Denotes the amplification over a bandwidth from 100 Hz to 3.5 MHz. A rating of 8 dB is typical.

3) *Intermodulation distortion.* Denotes the amount of distortion that results from intermodulation of two or more frequencies. A typical "worst-case" rating is 20 percent.

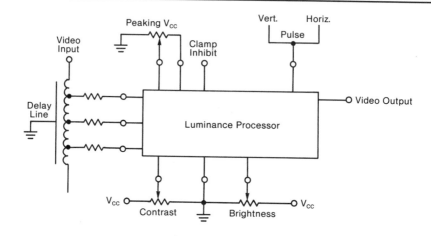

Figure 11-14

Input/output arrangement of an integrated circuit for luminance signal processing.

Consider next the functional arrangement of the composite video signal processing integrated circuit shown in Figure 11-15. This IC accepts a composite video signal from the first video amplifier and processes the signal to remove sync pulses, to generate tuner and IF AGC voltages, and to reduce the effects of possible interfering noise pulses. The RC network indicated in the diagram is a complex filter that operates between the output from the noise inverter and the sync separator.

A horizontal keying pulse is required for this circuit, as for the previously described video-IF section, which also generates its own AGC voltage. If such an IF section is used in the receiver, the AGC capability of the signal processor IC is not then required. The separated sync pulses are available in both positive and negative polarity. Tuner AGC voltage is also available in a bipolar output. Basic functional parameters for the IC are:

1) *Maximum power dissipation.* The exemplified IC can dissipate 750 mW at temperatures up to 55°C.

2) *Video input amplitude.* The nominal amplitude of the composite video signal has a typical rating of 3 p-p V.

3) *Sync output levels.* Maximum amplitude is equal to V_{cc}, with a typical value of 24 V.

4) *Horizontal pulse amplitude.* Typically ranges from 3 to 6 V.

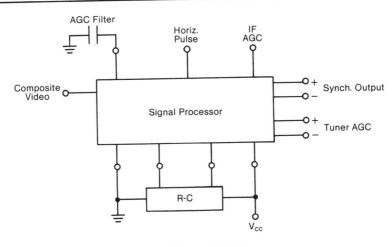

Figure 11-15

Input/output arrangement of an integrated circuit for composite video signal processing.

With reference to Figure 11-16, the functional arrangement of an integrated circuit for processing of the color signal is depicted. It is one of the most comprehensive color-signal processing ICs, and accepts the composite video signal from the video detector or from the first video amplifier of an IC. In addition, the chroma processor requires only a horizontal keying pulse to produce the color subcarrier reference signal, amplify the 3.58-MHz color subcarrier, demodulate the subcarrier, and produce the three color-difference signals.

This IC also contains an automatic color control (ACC) based on flesh tones, a color overload circuit, and a color-killer detector and amplifier. As shown in Figure 11-17, the automatic frequency and phase control (AFPC) requires an external filter, and a second filter is needed to remove the color subcarrier. The color reference oscillator is based on the 3.58-MHz external crystal and RC network. The two main color controls, the chroma gain and tint, are connected directly to the IC. An external switch permits the user to disable the automatic color correction system (ACC) and the overload protection circuit. Basic functional parameters for this IC are:

1) *Maximum power dissipation.* The maximum power that can be dissipated by the exemplified IC up to 55°C is 825 mW.

2) *Nominal power dissipation.* Denotes the total power dissipated under normal operating conditions. 500 mW is typical.

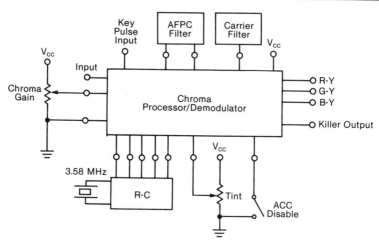

Figure 11-16

**Input/output arrangement of an integrated circuit
for processing the color signal.**

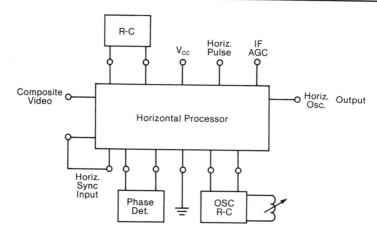

Figure 11-17

**Input/output arrangement of an integrated circuit
for a horizontal deflection signal processor.**

3) *Minimum oscillator pull-in range.* Denotes the maximum amount of mis-alignment of the crystal tuning capacitor with which the oscillator will still lock in phase with the reference signal. ±300 Hz is a typical value.

Next, consider the functional arrangement for an integrated circuit that processes the horizontal deflection signal, as shown in Figure 11-17. This IC contains the circuitry of a simple signal processor, plus horizontal automatic frequency control (AFC), horizontal oscillator, and a preamplifier for the horizontal sweep section. The IC separates the sync pulses from the composite video signal, provides some noise reduction, and also generates a keyed AGC signal. The RC section in the diagram contains the resistor and capacitor network required for the noise suppression function.

A separate external AGC filter is required. The horizontal and vertical sync-pulse output is connected to the horizontal sync input and to the vertical sweep section (which may be on another IC). An external phase-detector network, including resistors, capacitors, and two diodes, generates the error signal for the horizontal AFC. The horizontal oscillator itself depends on an external RC network; its frequency is adjustable by means of the tuned coil. An internal amplifier provides sufficient horizontal pulse output to drive the horizontal sweep and high-voltage section. Basic functional parameters for the IC are:

1) *Maximum power dissipation.* Up to 55°C, 750 mW is typical.

2) *Horizontal pulse amplitude.* A typical value of amplitude required is 25 p-p V.

3) *Sync signal output.* Denotes the amplitude of the vertical and horizontal sync signal. 1.5 p-p V is typical.

With reference to Figure 11-18 a functional arrangement for an integrated circuit that generates the horizontal and vertical sweep voltages is shown. This IC accepts the horizontal and vertical sync signals from a signal processor such as described above, and develops the horizontal and vertical sweep waveforms. An unusual technique is employed to ensure precise synchronization without vertical or horizontal hold controls. The horizontal flyback pulse input is compared with the horizontal sync input in a standard phase-detector circuit which controls the voltage-controlled oscillator (VCO).

An external loop filter is required to smooth out the error voltage. The 503.5-kHz ceramic resonator controlled-oscillator output is counted down in increments of 32 to generate the 15.75-kHz horizontal sweep signal. Twice the horizontal frequency is supplied to an internal timing circuit which also receives the vertical sync signal. The vertical RC network removes the horizontal pulses from the vertical sync pulses. The 60-Hz vertical sweep signal is always locked in with the horizontal sweep, ensuring correct interlace in the raster. The horizontal pulse is also used to provide a burst gating pulse which is applied to a color-processing IC for generation of the color sync burst. An internal regulator en-

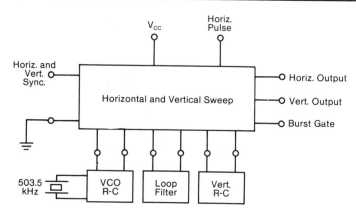

Figure 11-18

Input/output arrangement of an integrated circuit for generation of the horizontal and vertical sweep voltages.

sures stability of the power supply voltages. Basic functional parameters for the IC are:

1) *Maximum power dissipation.* Rated at 25°C; 830 mW is typical.

2) *Horizontal output frequency.* With no error voltage applied, the typical center frequency is 15,750 Hz.

3) *Horizontal output frequency range.* The normal range of output frequencies, controlled by the resonant 503.5-kHz resonant element, over which the oscillator can be varied by the error voltage is typically rated from 15,150 to 16,300 Hz.

4) *Vertical pull-in range.* The frequency range over which vertical sync pulses can be locked in is typically rated from 58.1 to 67.1 Hz.

5) *Horizontal pull-in range.* Frequency band over which the horizontal oscillator can be locked in; typically rated ±600 Hz.

6) *Horizontal static phase error.* With a frequency range of ±600 Hz, the typical phase error of the horizontal output signal is no more than ±0.5 microsecond.

VIDEO SWEEP MODULATION

Optimum evaluation of color signal-channel frequency response can be made by means of the video-sweep-modulation (VSM) technique as shown in Figure 11-19. This is a specialized type of alignment procedure that shows how

Figure 11-19

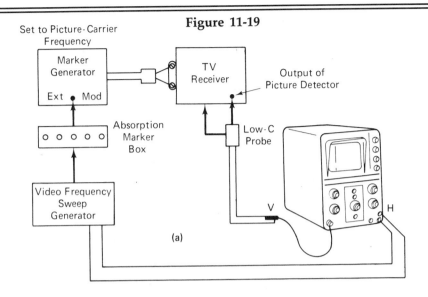

Set to Picture-Carrier
Frequency

Marker Generator

Ext • Mod

Absorption Marker Box

Video Frequency Sweep Generator

TV Receiver

Output of Picture Detector

Low-C Probe

V H

(a)

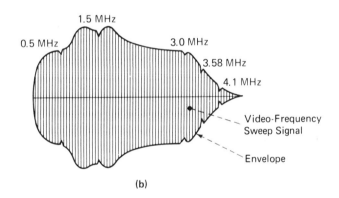

1.5 MHz

0.5 MHz   3.0 MHz

3.58 MHz

4.1 MHz

Video-Frequency Sweep Signal

Envelope

(b)

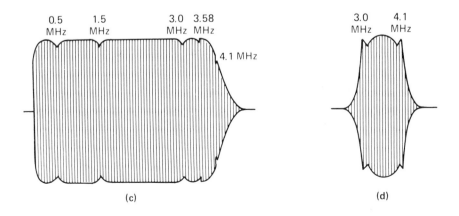

0.5 MHz   1.5 MHz   3.0 MHz 3.58 MHz

4.1 MHz

(c)

3.0 MHz   4.1 MHz

(d)

Figure 11-19 (Continued)

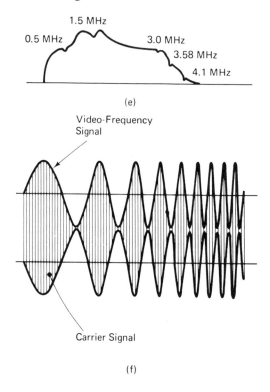

(e)

Video-Frequency
Signal

Carrier Signal

(f)

Figure 11-19

**Typical video-sweep-modulation test setup: (a)
equipment connections; (b) representative scope
pattern at picture-detector output; (c) typical
pattern at video-amplifier output; (d) pattern at
bandpass-amplifier output; (e) display of (b) when
a demodulator probe is used instead of a low-C
probe; (f) waveform of a video-sweep-modulation
signal.**

the various signal channels work together as a team. It employs an encoded
sweep signal. The VSM signal consists of a video-frequency sweep signal that is
modulated on the picture-carrier frequency. When the VSM signal is applied to
the input terminals of a TV tuner, the signal passes through to the IF amplifier
and is then demodulated by the picture detector. In turn, an oscilloscope connected at the output of the picture detector displays the combined RF-IF re-

sponse of the receiver, as exemplified in Figure 11-19(b). This is a video-frequency sweep signal that has been varied in amplitude according to the response curves of the front end and the IF strip. It has five absorption markers along its envelope, indicating key frequencies in the pattern. Next, if the oscilloscope is connected at the output of the video amplifier, the combined response of the RF, IF, and video amplifiers is displayed. A typical pattern is shown in Figure 11-19(c). Or, if the oscilloscope is connected at the output of the bandpass amplifier, the combined response of the RF, IF, video, and bandpass amplifiers is displayed, as seen in Figure 11-19(d). These patterns are obtained using a low-C probe; if a demodulator probe is employed, the pattern in (b) then appears as shown in (e). The waveform of a VSM test signal is depicted in (f).

TAPE RECORDER TROUBLESHOOTING

Basic Mechanical Difficulties • Tape Recorder Troubleshooting Chart • Overview • Microcassette Recorders • Tape Recorder Recycling Quick Check • Bias Voltage Considerations • Automatic Level Control • Related Factors • Variational Check • Frequency Distortion (Case History) • Hum Symptoms • Replacement Parts

BASIC MECHANICAL DIFFICULTIES

Many tape-recorder troubleshooting jobs are concerned with mechanical malfunctions. Functionally, a tape-drive assembly is employed to move the tape past the heads at a constant speed while the recorder is in operation. A rewind function is also provided in many machines which permits the tape to be reversed and rewound rapidly. Observe that while the tape is being recorded or played back, it is pressed against a rotating shaft (capstan) by a pinch roller. The capstan function is provided by a flywheel, and is driven by an electric motor. Note that a speed reduction arrangement (usually a belt-and-pulley assembly) is provided between the motor and capstan. (See Figure 12-1).

Defects in the mechanical system can cause trouble symptoms such as no tape movement, incorrect tape speed, faulty braking action, erratic operation, and riding of the tape up and down between capstan and pressure roller. Preliminary troubleshooting should start with a tape that is in good condition. Note that a tape that has been spliced may "foul up" between the capstan and the pressure roller. The capstan and pressure roller should be visually inspected and

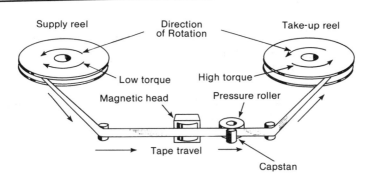

Reproduced by special permission of Reston Publishing Company, from Electronic Troubleshooting *by Clyde Herrick.*

Figure 12-1

Basic features of a tape drive assembly.

cleaned if there is evidence of grease, grime, or oxide on the working surfaces. Sometimes the capstan and/or roller will collect spots of adhesive from carelessly spliced tapes.

Observe that the pressure roller is provided with a spring. If the spring is weak or broken, tape-transport malfunction inevitably results. Worn or fouled belts also cause nonuniform movement of the tape. Erratic transport can also result from a dragging brake shoe that impedes forward transport. Adjust the brake mechanism, if necessary, to eliminate any drag. Also, check for the possibility of lack of lubrication and/or fouled bearings, both of which impose drag.

Note that faulty braking action is commonly caused by a weak or broken brake spring. Sometimes a brake shoe fails to retract because of loose screws. Braking failure will result if lubricating oil crawls over the braking surface. "Grabbing" is usually caused by gummy deposits and may be responsible for tape breakage. Observe that brake pads are provided in some designs; if the pads are worn out, they must be replaced.

When the tape tends to ride up and down between the capstan and the pressure roller, there is strong suspicion of lubricating oil on the capstan surface. Also, check the pressure-roller mounting stud to make certain that it is parallel to the capstan; if it is not parallel, it should be replaced with a new assembly. Another possibility of "riding" is a scratched or scored capstan; in such a case, the capstan must be replaced. Also, excessive take-up torque is a possible culprit—inspect the clutch and belt assembly for defects. Occasionally, a pressure roller becomes eccentric or flattened and in turn causes the tape to "ride" up and down. (See Table 12-1.)

OVERVIEW

A *tape recorder*, in the strict sense of the term, has both recording and playback functions. A *tape player* has a playback function only. A *tape deck* denotes any tape unit that has no power amplifier or speaker; it usually has no housing and is intended for custom installation in a high-fidelity system. A *record/play deck* records tapes for playback through a hi-fi system, in a car, or through a portable tape player. A *play deck* is designed to play car tapes or for "second systems."

MICROCASSETTE RECORDERS

We also encounter *microcassette recorders* which are designed as small mono units. Two tape speeds are typically provided: 2.4 cm/s and 1.2 cm/s. Some designs include voice-actuation facilities whereby the machine automatically switches on when there is microphone output, and then switches off when microphone output stops. Note that frequency response is somewhat limited—

Table 12-1
Tape Recorder Troubleshooting Chart

Trouble Symptom	*Most Probable Cause*
Completely "dead" machine.	Fuse may be blown; power ground might be defective; cartridge switch could be "open"; connection wires may be burned; circuit-board conductors may be burned.
No transport of tape; pilot lamp is illuminated.	Drive belt is often loose; belt may have run off pulley; pulley setscrew could have loosened; capstan drive may be frozen; look for broken motor power connection on circuit board.
Objectionable "wow" or flutter on playback.	Mechanism probably needs cleaning and lubrication; tape or cartridge may be defective; capstan could be defective or oily; drive belt might be loose or defective; motor sometimes binds and runs hot; pinch roller may be defective or frozen; look for a flat spot on the pinch roller; spring or rubber assembly might be frozen; capstan bearing could be dry or defective; speed-control circuit may be faulty.
Abnormal mechanical and/or electrical noise.	Inspect playback head, and demagnetize; check whether noise comes from L or R channel, or both; cartridge may be defective; motor bearings might be worn excessively; capstan sometimes freezes or seizes; noise may be produced by the ignition system; motor might be defective.
Weak or zero audio output on any or all tracks.	Speaker or wiring may be defective; tape is sometimes faulty; look for a deteriorated component in the amplifier; tape

(continued)

Table 12-1 (Continued)

Trouble Symptom	*Most Probable Cause*
	head might be defective or out of adjustment; inspect head for oxide deposits; if battery operated, check battery voltage; check for weak stage(s) by signal injection.
Crosstalk between tape tracks.	Head height needs adjustment; tape may be faulty; cam assembly might be defective; look for foreign substances in cartridge opening.
Poor high-frequency response.	Inspect head for oxide deposits; head may be excessively worn; head may be misaligned with respect to tape; check amplifier for a component defect.
Machine fails to change tracks automatically.	Solenoid circuit may be "open" or "shorted"; selector switch could be "open" or "shorted"; pawl sometimes becomes defective; cam mechanism might be faulty; check winding on solenoid for "open" or "short"; head index cam sometimes becomes frozen.
Tracks cannot be changed manually.	Manual switch may have failed; check switch contacts for corrosion or pitting; wires to switch might be broken; check to determine whether indicator arm is binding.
Fuse blows whenever track changes.	Diode shunting solenoid winding is probably defective.
Distorted audio output.	Speaker may be damaged or defective; tape may be faulty; check for a deteriorated amplifier component; measure the dc supply voltage; distortion can be

(continued)

Table 12-1 (Continued)

Trouble Symptom	Most Probable Cause
	caused by excessive supply voltage; look for a defective speaker connection.
Audio output is unbalanced.	Balance control may need adjustment; tape head might be faulty; check for speaker defects; height adjustment might be needed; check audio section for a weak stage.
Tape transport slows down.	Flywheel may be binding; belt might be slipping; inspect capstan bearing for wear or lack of lubrication; oil may have crept upon the belt or the capstan drive; motor could be defective; pulley might have worked loose.

Reproduced by special permission of Reston Publishing Co., Inc., and Derek Cameron from *ADVANCED ELECTRONIC TROUBLESHOOTING*.

typically, high-frequency response extends to 4 kHz. Some designs include a fast-play mode that transports the tape approximately 25 percent faster than its normal speed.

As a practical note, excessive tape slack inside the microcassette can cause malfunctions. Accordingly, it is good practice to make certain that any tape slack is taken up before loading a microcassette. To take up tape slack, insert a pencil into the core of the reel and gently turn the pencil to remove any slack.

Although the troubleshooter is usually concerned with cassette tape units, reel-to-reel machines are encountered on occasion. Basic troubleshooting procedures are essentially the same for both types of machines. A block diagram is shown in Figure 12-2, with principal features noted. Careful distinction should be made between electronic malfunctions and mechanical problems in preliminary troubleshooting procedures. For example:

1. Weak or zero output on all tracks can be caused by a faulty tape; or, the tape head may be defective or out of adjustment; the head should be inspected for oxide deposits.

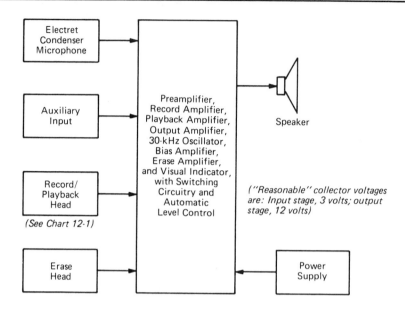

Note: When troubleshooting without service data, it is helpful to recognize the standard configurations and to visualize their functional sections. This enables rapid selection of meaningful quick checks with identification of the relevant test points.

Although the troubleshooter may not have a full understanding of the circuit sections and functions of a particular machine, its stereo design permits effective "easter egging" by comparative voltage, resistance, and impedance checks.

Smaller designs of cassette tape recorders employ simplified circuitry in which the playback amplifier also functions as the record amplifier. Similarly, smaller designs utilize a combination record/playback head, whereas separate heads may be provided in highly sophisticated designs.

Figure 12-2

Block diagram for a cassette tape recorder; nearly all tape machines are stereo designs.

2. Poor high-frequency response may be caused by an excessively worn head; the head should be inspected for oxide deposits; head alignment with respect to the tape should be checked.

3. Distorted audio output may be caused by a faulty tape; the speaker may be damaged or defective; batteries may be weak. In the absence of a mechanical problem, these three problems will be localized in the electronic circuitry.

4. Cross-talk between tape tracks is always caused by incorrect head adjustment, by a faulty tape, or by foreign substances in the cartridge opening.

5. Noisy output is likely to be caused by a magnetized playback head or by a faulty tape. An erratic tape transport can introduce noise. In the absence of a mechanical problem, however, noisy output will be localized in the electronic circuitry. For example, a transistor or resistor in the preamplifier may generate noise.

TAPE RECORDER RECYCLING QUICK CHECK

Distortion problems can be a headache compared with most of the catastrophic routine failures in tape machines, due to difficulties in evaluating distortion trouble symptoms. When distorted output occurs, the troubleshooter usually checks the tape first. In other words, if a chromium-dioxide tape is used on a machine designed for ferromagnetic tape, both recording and reproduction will be impaired. Another basic type of distortion results from cross-talk between tape tracks. In this situation, the troubleshooter checks out the playback head height, inspects the tape for possible damage, and looks for foreign substances in the cartridge opening.

If distortion involves poor high-frequency response, the troubleshooter inspects the playback head for oxide deposits, for excessive wear, or for misalignment. In case no mechanical faults are found, the amplifier should be checked for normal operation (preferably on a comparison basis). Note that if distorted output is accompanied by scraping, rattling, or buzzing sounds, the defect may be found in either the amplifier or the speaker. Loud popping sounds throw suspicion on the speaker voice-coil connections.

A tape-recorder recycling quick check, as shown in Figure 12-3, can be helpful in troubleshooting distortion trouble symptoms. This quick check consists merely of rerecording and replaying the same musical passage several times via a hi-fi tape recorder. In turn, the number of times that the recording can be recycled without intolerable distortion is a figure of merit for the tape recorder under test. Observe that as sound-track deterioration progresses, the most serious distortion factor becomes greatly exaggerated and therefore more easily identified.

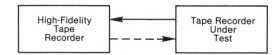

Note: This recycling quick test serves to exaggerate the most serious distortion factor and thereby facilitates analysis and fault identification. For example, after several recycling periods, the troubleshooter might clearly recognize "the sound of class B," or "the sound of clipping," or "the sound of subnormal bandwidth."

Figure 12-3

Tape recorder recycling quick check.

BIAS VOLTAGE CONSIDERATIONS

Since the oxide coating on a tape has a nonlinear magnetic characteristic, its transfer characteristic must be linearized by means of an AC bias voltage mixed with the audio signal. This AC bias voltage has a frequency in the 30 to 60 kHz range. Troubleshooting of the bias system involves the following considerations:

- *Optimum Amplitude:* The bias voltage should be adjusted to optimum amplitude, although the bias-oscillator frequency is not critical.
- *Frequency Response:* Note that the bias-voltage amplitude affects the frequency response of the transfer characteristic, as well as its linearity.
- *Output Level:* The bias-voltage amplitude also affects the output level of the recorded tape on playback.
- *Peak Bias:* The peak-bias adjustment corresponds to a bias-voltage amplitude that provides maximum output level on playback.
- *Overbias:* Maximum linearization (minimum harmonic distortion) is obtained by overbias—approximately 2 dB over peak bias.
- *Low-Frequency Response:* Optimum low-frequency response is obtained at peak-bias amplitude.
- *High-Frequency Response:* High audio-frequency response is reduced by overbias.
- *Underbias Effects:* Underbias adjustment results in distorted output, low signal-to-noise ratio, and reduced output level.

Thus, the AC bias-voltage adjustment involves certain conflicting factors, and the optimum value is necessarily a compromise.

AUTOMATIC LEVEL CONTROL

Cassette-deck and other tape machines often include automatic level control (ALC) facilities, as shown in Figure 12-4. ALC action prevents tape saturation and resultant distortion of musical passages that have a wide dynamic range. ALC also has the advantage of maintaining a reasonably uniform audio output, although various speakers may be situated at different distances from the microphone during recording of conversations. Many of the tape machines that employ ALC provide a switch for defeat of the ALC function if the operator so desires. Some tape machines include amplified ALC, which permits processing of a very extensive dynamic range without tape saturation by progressively compressing the higher amplitude peaks in the audio signal.

Note, however, that wide-range ALC operation has certain disadvan-

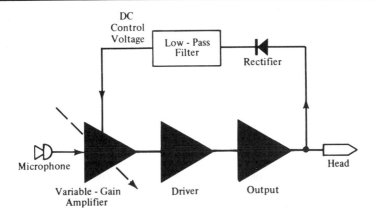

Note: An automatic level control system basically consists of a rectifier and a low-pass filter (integrating circuit) connected in a feedback loop from output to input of the amplifier. An ALC amplifier may be included in the loop. The ALC system functions to control the gain of the input stage in essentially the same manner as an automatic gain control system, or an automatic volume control system. The chief distinction from the troubleshooter's viewpoint is that an ALC system normally has a much longer time-constant than an AVC or AGC system.

Reproduced by special permission of Reston Publishing Company, and Lloyd Hardin from Advanced Stereo System Equipment.

Figure 12-4

Plan of an automatic level control arrangement.

tages. For example, when used inexpertly in problem environments, wide-range ALC can introduce objectionable distortion and noise. Another problem is that highly compressed musical passages tend to lose their "live" ambience. Observe that wide-range ALC systems, in particular, must be operated at a suitable preset level to accommodate the environment. Otherwise, background noise will rise and fall obtrusively as the ALC voltage changes from a high value to a low value (termed the "pumping" effect). A related "breathing" effect in re-production of musical passages is sometimes (incorrectly) regarded as an ampli-fier fault.

RELATED FACTORS

Elaborate cassette decks often provide a front-panel adjustable bias con-trol, whereby the operator can optimize the sound output for conventional, chrome, or metal tape. The maximum bias-voltage level is typically 0.7 r.m.s. volt. Note that poor high-frequency response can be caused by a reversed tape, as well as by an excessively high bias voltage. A magnetized recording head, defective microphone, or even an incorrect setting of the tone control can also cause poor high-frequency response.

Although less common than mechanical or amplifier defects, a distorted bias waveform can also cause noisy and/or distorted sound reproduction. Wave-form distortion results from a faulty device or component in the bias-oscillator section.

Hum interference in the output points to defective ground connections— check the braid connection on the microphone cable. In line-operated equip-ment, the power-supply filter section is a prime suspect. Some decks are provided with polarized power plugs—if the broad prong is "trimmed" or otherwise defeated, audible hum may result from reversed insertion of the plug into a power outlet.

VARIATIONAL CHECK

A variational check for audio stage distortion is shown in Figure 12-5. An audio generator is connected to the input terminals, and a DC voltmeter is connected at the output terminals of the stage. To check for distortion, the generator output is increased from zero to the maximum rated input level for the stage. As the signal level is varied, the DC voltmeter reading is observed. *If any shift occurs in the voltmeter reading, the troubleshooter concludes that the stage is operating in a nonlinear manner.*

The principle of this variational check for distortion is based on the partial rectification that occurs when a signal passes through a stage that has a nonlinear transfer characteristic. Compression of the peak signal excursion re-sults in more or less rectified signal current flow through the collector load resistor. In turn, the average DC in the collector circuit changes, and a shift in collector voltage occurs.

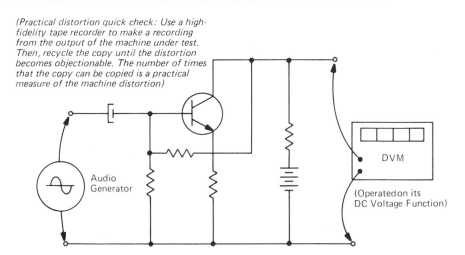

(Practical distortion quick check: Use a high-
fidelity tape recorder to make a recording
from the output of the machine under test.
Then, recycle the copy until the distortion
becomes objectionable. The number of times
that the copy can be copied is a practical
measure of the machine distortion)

Note: If the stage is operating in class A at maximum rated power output, there will be no shift in DC level from the minimum to the maximum drive condition. A shift in DC level indicates that the transistor is operating in part as a rectifier (nonlinearly). As a cautionary note, this is not a completely informative test. If the transistor happens to compress or to clip both the positive peaks and the negative peaks of the sine wave by the same amount, there will be no shift in DC level, although distortion is taking place.

As another cautionary note, an oscilloscope has limited ability to provide adequate indication of small percentages of distortion. For example, it is difficult to see 2-percent distortion in a displayed sine wave.

This test setup is also used to check the frequency response of an audio amplifier. However, the DVM is then operated on its AC voltage function, with a series blocking capacitor.

Figure 12-5

**Test setup for variational analysis of stage
distortion.**

Note that a variational check is not conclusive proof that a stage is free from distortion. In other words, if a shift in collector voltage occurs, the troubleshooter knows that the stage is distorting. On the other hand, if no shift in collector voltage occurs, the stage may nevertheless distort the signal. If both the positive peak and the negative peak of the signal happen to be equally compressed, no shift in DC collector voltage will occur, although the signal is being distorted.

FREQUENCY DISTORTION (CASE HISTORY)

A nonlinear transfer characteristic causes amplitude (harmonic) distortion. Frequency distortion, on the other hand, is generally caused by defective capacitors. For example, consider the case history depicted in Figure 12-6. The sound output had a "tinny" quality and lacked lower audio frequencies. DC voltage values were normal, and a low-power ohms check showed that the resistive values were within tolerance. A bias-on test indicated that the transistor was workable.

Then, it was recognized that the 0.0015 μF capacitor in the negative-feedback circuit from collector to base functions as a frequency-responsive component (provides increasing negative feedback at higher audio frequencies). Accordingly, the troubleshooter checked the feedback capacitor and found that it was open. Replacement of the capacitor restored the amplifier to normal frequency response.

Figure 12-6

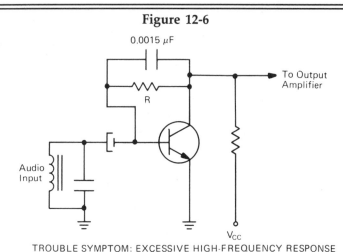

TROUBLE SYMPTOM: EXCESSIVE HIGH-FREQUENCY RESPONSE

Note: Frequency distortion occurs when the 0.0015 μF feedback capacitor is open.

This is an example of a frequency-selective negative-feedback arrangement. In other words, the reactance of the feedback capacitor decreases as the frequency increases, thereby developing less stage gain at higher frequencies.

Figure 12-6 (Continued)

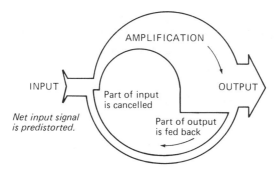

Note that resistor R also provides negative feedback from collector to base. This negative feedback component, however, is not frequency-selective, inasmuch as the resistor has the same value at all audio frequencies.

Figure 12-6

Example of a frequency-compensated stage. Case history of distortion caused by an open capacitor.

HUM SYMPTOMS

Low-level circuits are much more susceptible to hum and interference pickup than are high-level circuits. An audio system can be divided in a general way into low-level circuits and high-level circuits. For example, the output circuit from a tape recorder is a low-level circuit. On the other hand, the output circuit from a preamplifier is a comparatively high-level circuit.

Most of the interconnections in an audio system are made with audio cable. This is a shielded and stranded type of conductor; the shield braid that surrounds the central conductor is grounded to the chassis (or to the common supply bus) of the audio equipment. Therefore, undesirable pickup of hum and various other stray fields is virtually eliminated.

Occasionally, in spite of observance of all good practices in installation procedures, an audio system may pick up objectionable interference. This trouble symptom results from operation in a location that has abnormally high stray-field intensity. In turn, the low-level circuits must be very well shielded, and the braid of the associated audio cables must be well grounded. In addition, it is good practice to connect a separate ground wire from a record player to its preamplifier. This should be a heavy copper wire to provide a low-impedance ground return for the phono circuit, and thereby minimize the possibility of audible hum from this source.

REPLACEMENT PARTS

After a fault has been localized in a tape recorder, the troubleshooter must then obtain suitable replacement parts or devices. Procurement of replacement parts can sometimes be a headache. Electronics parts jobbers occasionally have difficulty in identifying a particular part, or even if it can be identified, in finding a suitable supplier. However, a resourceful troubleshooter need not be defeated in this situation—if a bit of time is spent in looking up a similar cast-off tape recorder, it is quite likely that the needed replacement part can be salvaged from the cast-off unit. For example, thrift shops and good-will stores generally have boxes of cast-off tape recorders that "need work," and can be purchased for a trivial sum.

13

CLOSED CIRCUIT TELEVISION CAMERA TROUBLESHOOTING

Overview • Preliminary Trouble Analysis • AC-Coupled Video-Signal Output • Pinpointing Defective Parts • Simulated Vertical Sync Pulse • Differentiation and Integration of Basic Waveforms • Recognition of Nonlinear (Amplitude) Distortion • Blanking Circuitry • Video Amplifier • VHF Modulator • Waveform Distortion and Negative Feedback • Vidicon Section

OVERVIEW

A closed circuit television camera (CCTV) is essentially a miniature television transmitter. It generates television signals that are not broadcast, but rather transmitted over a closed circuit and received only by interconnected receivers. Most CCTV cameras provide modulated RF output on one or two VHF channels, plus video-frequency output for reproduction on a monitor unit. Thus, if the trouble symptom is "no VHF output," the camera should be checked for video output—in case the video output is normal, the trouble will be found in the RF oscillator/modulator section.

When a CCTV camera is workable, and the trouble symptom is "poor picture," picture analysis can frequently provide helpful clues concerning the fault and its probable location. A standard test pattern such as shown in Figure 13-1 is preferred, although useful information can be obtained by analyzing the reproduction of any scene, such as the interior of a shop or living room. The troubleshooter is concerned with factors such as image proportions, scanning linearity, brightness, contrast, picture detail, distortion, noise, and interference.

Troubleshooting of CCTV circuitry without service data is accomplished to best advantage by means of comparison tests with respect to a similar camera that is in normal operating condition. However, a comparison camera may not be available, and other approaches must then be used. Observe that the PC board in a CCTV camera may be marked in various ways. For example, parts numbers may be indicated. When parts numbers are not indicated, the PC board may still have several key test points marked. Note that the oscilloscope is second in usefulness only to the DVM in "buzzing out" CCTV circuitry. If a scope is not available, the troubleshooter must then rely on the DVM and various quick checks. Sometimes temperature tests will turn up an overheated transistor at the outset.

PRELIMINARY TROUBLE ANALYSIS

Closed-circuit television cameras use the basic sections depicted in Figure 13-2. A vidicon camera tube, deflection yoke, video amplifier, deflection circuits, sync waveshapers, and power supply are employed. The 1-inch vidicon camera tube forms the camera signal, which in turn drives the video amplifier. This video amplifier supplies a complete picture signal at 1 V p-p into a 75-ohm load.

Note that the camera signal, sync pulses, and blanking pulses are com-

Figure 13-1

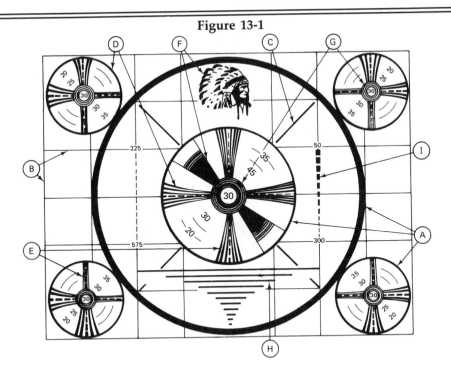

Note: Information contained in the test pattern includes:

A. Various portions of the test pattern are circular forms that indicate scanning linearity, adjustment of height and width controls, and picture centering.

B. Other portions of the test pattern are square forms that indicate the aspect ratio (normally 4 to 3). If the outline of the test pattern is not a true rectangle, keystone distortion is indicated. Reproduction of the "grid" pattern also displays whether vertical and/or horizontal scanning nonlinearity may be occurring.

C & D. Vertical resolution (reproduction of detail) is shown by the horizontal wedges. Note the point toward the center of the pattern at which the wedges start to look "blurry" and add a zero to the figure. For example, if the wedges start to become indistinct at "30," the vertical resolution is equal to 300 lines.

E. Horizontal resolution is shown by the vertical wedges. Note the point toward the center of the pattern at which the wedges start to look "blurry," and add a zero to the figure. For example, if the wedges become indistinct at "30," the

Figure 13-1 (Continued)

horizontal resolution is equal to 300 lines. Next, to find the picture-channel bandwidth, divide the horizontal resolution figure by 80; in the foregoing example, 300/80 = 3.75 MHz, approximately.

F. Linearity of amplification in the picture channel is indicated by reproduction of the diagonal wedges. A succession of four shading tones is provided; these gray tones are normally reproduced in distinct steps.

G. Vidicon tube focus may be checked on the basis of reproduction of the center circles, or "bull's-eyes."

Reproduced by special permission of Reston Publishing Company, and Derek Cameron from Advanced Electronic Troubleshooting.

Figure 13-1

A standard test pattern for analysis of circuit actions and malfunctions.

bined in the video amplifier; sync action utilizes random interlace in this example. The video amplifier drives the VHF oscillator and modulator section which normally provides an output level of 50,000 μV into a 300-ohm load on any channel from 2 to 5. Preliminary troubleshooting proceeds as follows:

CW Output, No Video. This is a comparatively common trouble symptom. It points to possible defects in the video amplifier, deflection section, or vidicon. *Case History:* The video signal was absent for 10 minutes after the CCTV camera was turned on; however, the VHF carrier signal was immediately outputted. The trouble was tracked down to a marginal electrolytic capacitor connected in series with the video-amplifier output.

Dead Camera. This is a less common trouble symptom. It points to power-supply failure, as detailed subsequently.

Video Output, No VHF Output. This trouble symptom points to a defect in the VHF oscillator and modulator section.

Camera Signal Weak. Check the vidicon terminal voltages; if ok, the tube should be checked by substitution.

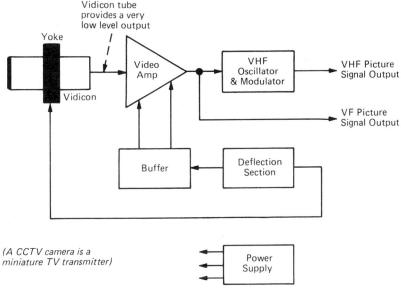

(A CCTV camera is a
miniature TV transmitter)

WHEN TROUBLESHOOTING WITHOUT SERVICE DATA,
AND WITHOUT AN IDENTICAL CAMERA FOR COMPARISON
TESTS, TRANSISTORS CAN BE "BUZZED OUT" AS EXPLAINED
IN CHART 1-1.
AN IN–CIRCUIT TRANSISTOR TESTER IS ALSO VERY
HELPFUL IN PRELIMINARY TROUBLESHOOTING
PROCEDURES.

Note: When there is no VHF picture signal output, and no video-frequency picture output, an audio-frequency test signal can be injected at the vidicon output to determine whether the trouble is in the camera circuitry or in the vidicon tube.

Troubleshooting without service data is greatly facilitated by the availability of an identical camera in normal operating condition for making comparative voltage, resistance, and impedance checks. If comparative checks are not feasible, the troubleshooter is thrown upon his or her own resources. In this situation, a comprehensive understanding of standard configurations and sectional functions is very helpful, so that the troubleshooter can evaluate test measurements meaningfully.

Figure 13-2

Block diagram for a typical CCTV camera.

AC-COUPLED VIDEO-SIGNAL OUTPUT

The video-signal output in standard CCTV cameras is rated at 1 volt peak to peak. This voltage can be checked with an oscilloscope, or with a peak-to-peak DVM. Troubleshooters should keep in mind that the video-signal output is usually AC-coupled to its source. This means that the average value of the video output signal is always zero, and that a DC voltmeter will normally read zero. However, a peak-reading DVM will provide a check of the video-signal amplitude; the peak voltage is always less than the peak-to-peak voltage. From the troubleshooter's viewpoint, an important fact is that the reading of a peak-reading DVM normally changes as the CCTV camera is focused first on a dark scene, and then on a light scene. If the reading does not change, there is a fault in the video channel. Note that the indication of an average-reading DVM also normally changes when this quick check is made. (See Figure 13-3).

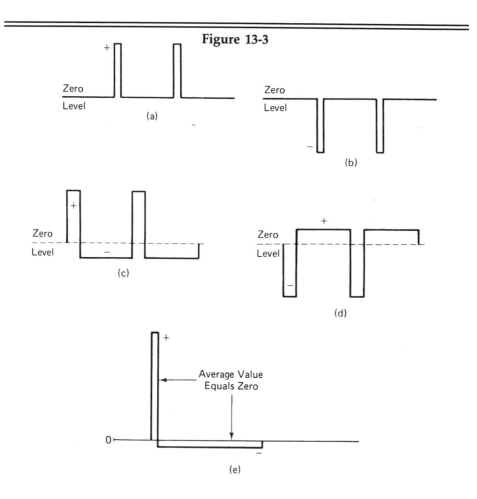

Figure 13-3

Figure 13-3 (Continued)

Note: The waveform at (a) is an example of a DC positive-pulse waveform; it has zero negative excursion. The waveform at (b) is an example of a DC negative-pulse waveform; it has zero positive excursion. If a DC pulse waveform is coupled through a capacitor, the output is changed into an AC pulse waveform. The waveform at (c) is an example of a positive-going AC pulse waveform; it has both positive and negative excursions. The waveform at (d) is an example of a negative-going AC pulse waveform; it has both positive and negative excursions. The waveform at (e) is an example of a positive-going AC pulse waveform, illustrating that its average value is zero (the areas above and below the zero-volt axis are equal).

Reproduced by special permission of Reston Publishing Company, and Douglas Bapton from Modern Oscilloscope Handbook.

Figure 13-3

Basic characteristics of DC and AC pulse waveforms.

PINPOINTING DEFECTIVE PARTS

From the troubleshooter's viewpoint, a CCTV camera consists of the following sections: power supply, focus-coil current regulator, vertical regulator, vertical deflection, horizontal deflection, blanking generator, video amplifier, and VHF oscillator-modulator sections.

Power Supply. Refer to Figure 13-4. First a DC voltage measurement should be made at point G. In this example, a reading of +18 volts is normally observed. An abnormal or a subnormal voltage value points to trouble in the power supply circuitry, or possibly to an excessive current demand by the camera circuitry.

In the case of power-supply trouble, the AC voltage at the power-line outlet should be measured. If the power-line voltage is normal, check next for a blown fuse (F1 in Figure 13-4). However, if the fuse is not blown, the next step is to remove the cover from the camera, plug in the line cord, turn on the power switch, and observe the vidicon tube to see whether its heater is glowing.

If the heater is dark, measure the heater supply voltage; this value is normally 6.3 V AC. If the heater supply voltage is correct, the heater is probably burned out; if the heater supply voltage is weak or zero, the power transformer is probably defective. *Note that even if the power-supply voltage is correct, operating trouble can be caused by excessive ripple*—measure the ripple voltage (if any) on the AC function of a DVM.

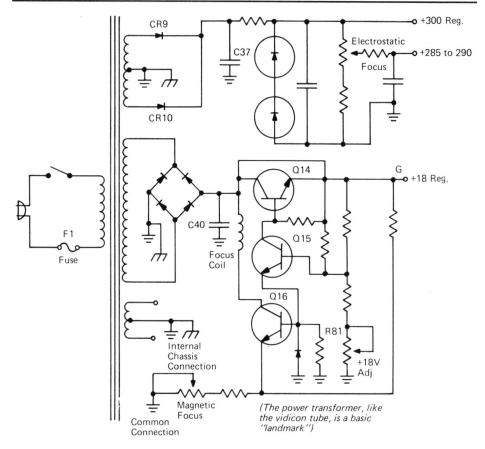

Note: The +18 V regulated output normally varies slightly with line-voltage fluctuation. For example, the following measurements are typical:

Line, 117.4 V, Reg, 18.06 V; Line, 110.1 V, Reg, 18.04 V; Line 100.0 V, Reg, 18.03 V; Line, 95.0 V, Reg, 18.02 V; Line, 90.1 V, Reg, 17.96 V; Line 85.0 V, Reg, 17.50 V.

Figure 13-4

Power-supply circuitry for the exemplified TV camera. If feasible, comparative temperature checks can be helpful.

SIMULATED VERTICAL SYNC PULSE

Troubleshooters should note that elaborate CCTV cameras generate standard vertical-sync pulses, whereas economy-type cameras generate simulated vertical-sync pulses, as shown in Figure 13-5. The composite video signals exemplified in parts (a) and (b) are normally present at the video-output terminal of a CCTV camera while scanning a test pattern. The vertical-sync interval in the composite video waveform is quite compressed, unless it is expanded on the scope screen, as shown in part (c). Observe that this standard waveform contains equalizing pulses following the horizontal sync pulses, and the vertical sync pulse per se is a serrated pulse. On the other hand, the simulated vertical-sync pulse dispenses with equalizing pulses and a serrated vertical-sync pulse. Instead, the simulated pulse is essentially a vertical-blanking pulse.

Technically, the distinction between standard vertical sync and simulated vertical sync is that the former provides for interlaced scanning action, whereas the latter is associated with random interlacing action. Random interlace involves a tradeoff between high-resolution images and circuit complexity. In other words, the horizontal oscillator is free-running during the vertical-sync interval when simulated vertical sync is utilized. Note that when standard vertical sync is used, a hammerhead pattern is formed during the vertical-sync interval, as shown in Figure 13-6. This hammerhead pattern becomes visible on the picture-tube screen when the vertical blanking pulse is disabled, the brightness control is advanced, and the picture is "rolled." Observe that when simulated vertical sync pulses are used, no hammerhead pattern is formed, and the vertical-sync interval is displayed as a solid black bar.

Power Transformer Circuit. To check the power-transformer circuitry, the CCTV camera's line cord is unplugged, and the resistance is measured between the prongs on the plug. With the power switch turned on, a reading of 65Ω is normally observed in this example. A reading of infinity can be caused by a *defective line cord,* a *defective switch,* a *blown fuse,* or a *burned-out primary winding.* A reading substantially less than 65Ω points to a partial short-circuit, such as short-circuited layers in the primary winding.

If the primary circuit checks out satisfactorily, the troubleshooter proceeds to measure the output voltages on the secondary side of the transformer. If one or more of the secondary voltages is subnormal, there is suspicion of short-circuited turns or layers in the secondary winding. (A zero reading usually results from an open-circuited winding.)

Next, if all of the secondary voltages measure correctly, the troubleshooter proceeds to make a DC voltage measurement at the positive terminal of C40 (Figure 13-4). A zero reading indicates that CR13 is probably defective. But if the measured value is normal (+26 V), the troubleshooter proceeds to point G and measures its voltage to ground (normally +18 V in this example). If

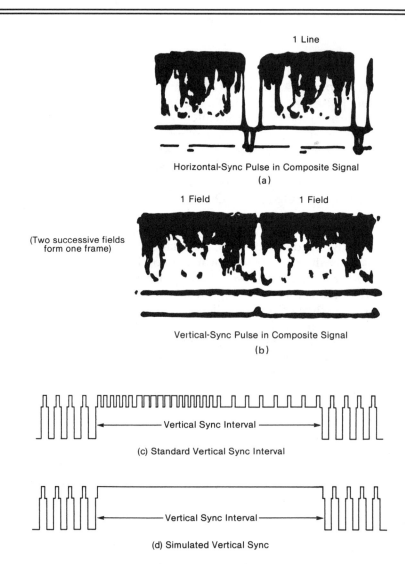

1 Line

Horizontal-Sync Pulse in Composite Signal
(a)

1 Field   1 Field

(Two successive fields
form one frame)

Vertical-Sync Pulse in Composite Signal
(b)

(c) Standard Vertical Sync Interval

Vertical Sync Interval

(d) Simulated Vertical Sync

Vertical Sync Interval

Reproduced by special permission of Reston Publishing Company, and Derek Cameron from Advanced Electronic Troubleshooting.

Figure 13-5

Composite video signal patterns.

Note: This is a useful quick check because it shows at a glance whether the generated sync waveform is normal. Observe that the hammerhead is invisible or only slightly visible at the normal setting of the brightness control. As the brightness control is advanced, the white level becomes excessively bright, the black level becomes gray, and the hammerhead is then plainly visible as a black image on a gray background. If a normal hammerhead pattern is displayed, the troubleshooter does not need to make an oscilloscope check of the generated sync waveform.

Reproduced by special permission of Reston Publishing Company, and Douglas Bapton from Modern Oscilloscope Handbook.

Figure 13-6

Normal appearance of the hammerhead pattern on the picture-tube screen.

this reading is zero, Q14 will probably be found defective; or if the reading is the same as at the positive terminal of C40, Q14 is probably short-circuited.

If the voltage at point G cannot be set to +18 volts by adjustment of R81, the remaining voltages in the regulator circuit should be measured. Specified DC voltages for this circuit are included in Table 13-1, for the exemplified circuitry.

Table 13-1
Specified Device Terminal Voltages

Transistor	Emitter	Base	Collector
Q1	+ 3.7 V DC	+ 3.35 V DC	+15 V DC
Q2	0 (zero)	+ 0.55	+ 4.6
Q3	+ 4.0	+ 4.6	+ 8.8
Q4	0 (zero)	+ 0.65	+12.5
Q5	+ 0.75	+ 1.15	+11.0
Q6	+ 5.5	+ 6.2	+12.0
Q7	+11.5	+12.0	+18.0
Q8	0 (zero)	+ 0.3	+ 3.0
Q9	0 (zero)	+ 0.65	+ 1.2
Q10		(See Text)	
Q11		(See Text)	
Q12	0 (zero)	− 1.8	+ 8.4
Q13	+ 0.4	+ 1.1	+ 7.3
Q14	+18	+18.5	+26
Q15	+ 6	+ 6.6	+18.5
Q16	+ 5.4	+ 6	+13

Next, if the foregoing tests clear the regulated power supply from suspicion, the technician turns his attention to the 300-V supply in this example. Note that a DC voltmeter cannot show whether the voltage is pure DC or pulsating DC (DC with an AC ripple component). The voltage at the positive terminal of C37 in Figure 13-4 is normally 340 V. A zero reading indicates that CR9 and CR10 are probably defective; a short-circuit fault could also be present.

On the other hand, if all of the power-supply voltages measure specified value, the troubleshooter concludes that the malfunction is not in the power supply, and proceeds to the next logical probability.

Horizontal Deflection. An out-of-sync trouble symptom results if the oscillator in the horizontal-deflection section (Figure 13-7) runs too fast or too slowly to lock in on the associated TV receiver or video monitor. R56 is adjusted as required—however, if R56 is out of range, Q10 will probably be found defective, in this example. If the pulse amplitude measures incorrectly, (normally, 65 V p-p, 10μs width, 15,750 Hz repetition rate), R58 is adjusted as required. However, if R58 is out of range, Q11 will probably be found defective. Normal DC voltages on Q10 and Q11 are:

Q10: emitter, +7 V; base 1, +0.6 V; base 2, +16 V.
Q11: emitter, +8.8 V; base, +8.8 V; collector, +4.2 V.

A sawtooth waveform with an amplitude of approximately 7 V p-p is normally present at the emitter of Q10; it can be checked as previously explained. If this waveform is absent, Q10 is not oscillating. A positive pulse with a width of 10 μs and an amplitude of approximately 3 V p-p is normally present at base 2 of Q10. If this waveform is missing, Q10 is probably defective. The deflection yoke and coils L5 and L6 are checked by resistance measurements.

Figure 13-7

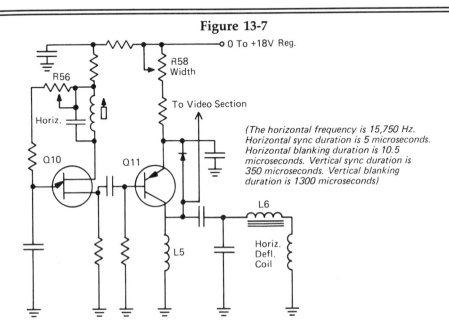

(The horizontal frequency is 15,750 Hz. Horizontal sync duration is 5 microseconds. Horizontal blanking duration is 10.5 microseconds. Vertical sync duration is 350 microseconds. Vertical blanking duration is 1300 microseconds)

Caution: When troubleshooting the horizontal-deflection section, be careful not to underscan by setting R58 to a low resistance. This error could cause a raster burn.

Note: Specified voltage values (see text) are with reference to the +18 V regulated supply. Check the supply voltage and adjust to 18 V, if required, as explained under "Power Supply."

An oscilloscope is the most useful instrument for identifying the various circuit sections, provided that the troubleshooter recognizes the basic types of waveforms that are encountered in TV circuitry. When troubleshooting without an identical TV camera for comparison tests, the troubleshooter also needs to identify waveforms that have been subjected to various forms of distortion. With adequate knowledge of circuit action, an educated guess can be made concerning the circuit fault that is causing the waveform distortion.

Figure 13-7 (Continued)

Reproduced by special permission of Reston Publishing Co., and Robert Russell from Electronic Troubleshooting with the Oscilloscope.

Figure 13-7

Horizontal-deflection circuitry for the exemplified camera. If feasible, comparative temperature checks can be helpful.

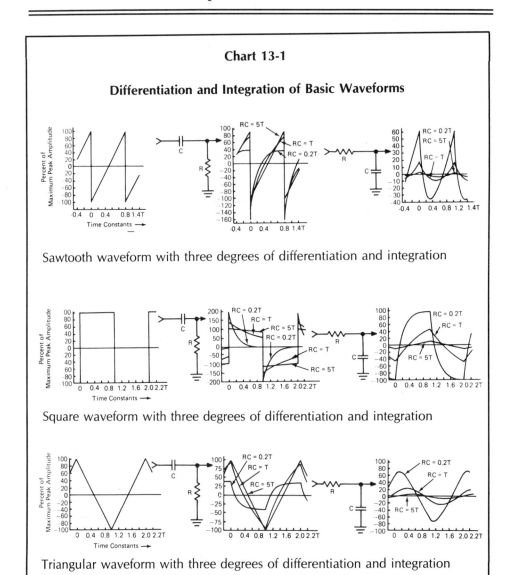

Chart 13-1

Differentiation and Integration of Basic Waveforms

Sawtooth waveform with three degrees of differentiation and integration

Square waveform with three degrees of differentiation and integration

Triangular waveform with three degrees of differentiation and integration

DIFFERENTIATION AND INTEGRATION OF BASIC WAVEFORMS

Waveforms become distorted in malfunctioning circuits, and the distortion features are related to the type of malfunction. With reference to Chart 13-1, it is seen that when the time-constant of a coupling (RC) circuit is too short, a sawtooth waveform exhibits ramp curvature. The amount of curvature that is introduced depends on how much the normal time-constant has been reduced. Since the flyback interval of a sawtooth waveform has a much more rapid rate of change than the ramp interval, very little curvature is introduced into the flyback trace.

If a sawtooth waveform is passed through an integrating circuit with a comparatively long time-constant, both the ramp interval and the flyback interval exhibit curvature. Again, the amount of curvature that is introduced depends on how much the time-constant has been increased. Observe that the distortions shown in Chart 13-1 are examples of frequency distortion. In other words, the malfunctioning circuits are still linear circuits. In case nonlinear distortion accompanies frequency distortion, particular regions of the frequency-distorted waveforms will be further distorted by compression and/or expansion.

Vertical Deflection. An out-of-sync trouble symptom results if the oscillator in the vertical-deflection section runs too fast or too slow. An overall check of the vertical-deflection section exemplified in Figure 13-8 can be made by

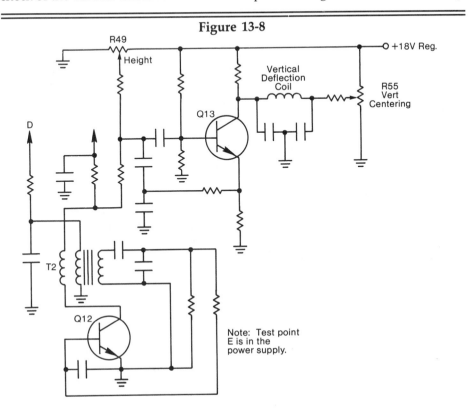

Figure 13-8

Figure 13-8 (Continued)

Caution: When troubleshooting in the vertical-deflection section, be careful not to underscan by setting R49 to a low resistance. This error could result in raster burn.

Note: When comparative checks can be made on an identical camera which is in normal operating condition, temperature checks are very helpful, as explained in Chapter 1. Open capacitors, or capacitors with a poor power factor can be pinpointed with an impedance checker, as explained in Chapter 2.

Vertical-deflection circuitry can be spotted with an aural signal tracer, such as shown in Figure 2-1. The vertical-oscillator transformer is a helpful landmark.

Figure 13-8

Typical vertical-deflection circuitry. If feasible, comparative temperature checks can be helpful.

waveform measurement at the collector of Q13. Normally, a negative-going sawtooth with a 60-Hz repetition rate is present; it can be checked with an oscilloscope. *If this waveform is weak, absent, or distorted, the troubleshooter proceeds to check the drive waveform into Q13.*

This is normally a positive sawtooth with an amplitude of approximately 0.5 peak-to-peak volt. Since its amplitude is comparatively low, a waveform check should be made with a preamp. The troubleshooter will also find it helpful to check the waveform at the collector of Q12. This is normally a negative-going pulse with an amplitude of approximately 7 V p-p, and a width of about 1.3 m.s. In case of zero output from Q12, check for a base driving waveform—if this waveform is missing, check for AC input at test points D and E, and at the primary of the blocking transformer T2.

To pinpoint a defective device or component, follow-up DC voltage measurements are made which may be supplemented by resistance measurements. Normal DC voltages for this example are listed in Table 13-1. Note that the vertical sweep amplitude is adjusted by setting R49 for an aspect ratio of 4 to 3 on the screen of the associated TV receiver or monitor. Vertical centering is adjusted by setting R55 to a point such that the DC voltages on either side of the vertical-deflection coil to ground are equal.

SYNCHRONIZING CIRCUITRY

Out-of-sync trouble symptoms can also be caused by malfunctions in the synchronizing section. With reference to the example in Figure 13-9, the troubleshooter usually starts by measuring the DC voltage values at the transistor

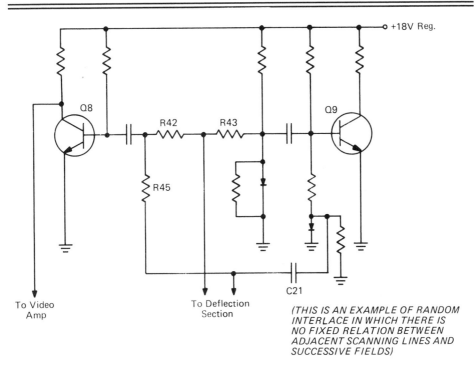

To Video
Amp

To Deflection
Section

*(THIS IS AN EXAMPLE OF RANDOM
INTERLACE IN WHICH THERE IS
NO FIXED RELATION BETWEEN
ADJACENT SCANNING LINES AND
SUCCESSIVE FIELDS)*

Note: The transistors can be quick-checked by means of a shut-off test. Apply a short circuit temporarily between the base and emitter terminals. In turn, the collector voltage will "jump up" to ±18 V if the transistor is in normal condition.

To quick-check Q8, the collector output conductor to the video amp must be temporarily open-circuited—a razor slit may be made to open the conductor, and subsequently closed with a small drop of solder.

Figure 13-9

Synchronizing circuitry for the exemplified TV camera. The vertical frequency is power-line locked at 60 Hz.

terminals. Specified voltages are listed in Table 13-1. Follow-up resistance measurements may be made with a low-power ohmmeter. If the troubleshooter needs additional data, he or she may proceed by checking out the combined sync waveform at the collector of Q9. This is normally a mixture of horizontal sync pulses and vertical sync pulses. Their presence or absence can be verified by "beating out" 60-Hz and 15,750-Hz frequencies with an audio oscillator as explained previously.

1. If no 60-Hz component is found, the troubleshooter proceeds to check for the vertical-frequency input signal at the junction of R45 and C21.
2. On the other hand, if no 15,750-Hz component is found, the troubleshooter checks for the horizontal-frequency input signal at the junction of R42 and R43.

A check for the combined sync waveform can also be made at the base of Q9. This waveform normally has an amplitude of approximately 1.8 V p-p, in this example. Leaky capacitors are the most common culprits, followed by open capacitors. *Caution:* If a diode is replaced, be careful to observe correct polarity. *If a replacement diode is installed with reverse polarity, the troubleshooter is likely to have a tough-dog situation to contend with.*

RECOGNITION OF NONLINEAR (AMPLITUDE) DISTORTION

Nonlinear distortion characteristics are different from frequency distortion characteristics. As a distinctive example shows, a basic sine wave is affected quite differently by passage through a nonlinear circuit, and by passage through a differentiating circuit. Thus, when a sine wave passes through a diffentiating circuit, its waveshape is unchanged; however, the output waveform is advanced in time with respect to the input waveform. (The differentiating circuit advances the phase of the sine waveform.) On the other hand, when a sine wave passes through a circuit with a nonlinear amplitude characteristic, its waveshape is changed, but its phase is unaffected.

With reference to Figure 13-10, the sine wave is applied to a compression circuit which progressively reduces the output amplitude in the positive-peak region. The result of nonlinear (amplitude) distortion is virtually invisible for low percentages of distortion. However, the resultant change in waveshape is quite obvious for high percentages of distortion. In malfunctioning circuitry, the troubleshooter may encounter a combination of nonlinear distortion and frequency distortion. Analysis of such distorted waveforms is facilitated by a clear recognition of basic distortion characteristics.

BLANKING CIRCUITRY

Visible retrace lines in the reproduced picture point to a malfunction in the blanking circuitry. With reference to the example of Figure 13-9, the first step is to measure the DC voltage at the terminals of transistors Q8 and Q9. (See Table 13-1 for specified voltage values in this example.) Follow-up resistance measurements may be made with a low-power ohmmeter.

Then, if additional test data are needed, the troubleshooter should check

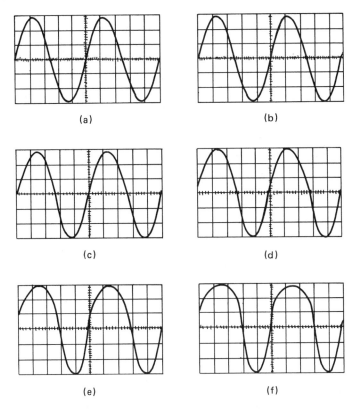

Note: These distorted sine waves show the effect of progressive amounts of compression in the positive-peak region of the waveform. The percentage of distortion in (a) is 1 percent; in (b), 3 percent; in (c), 5 percent; in (d), 10 percent; in (e) 15 percent; in (f), 20 percent.

Figure 13-10

Basical example of nonlinear (amplitude) waveform distortion.

out the combined blanking waveform at the collector of Q9, as explained above. If the vertical pulse (60-Hz fundamental component) is absent, the trouble-shooter will check next for the vertical waveform at the junction of R45 and C21. On the other hand, if the vertical pulse (15,750-Hz component) is missing, the troubleshooter will check next for the horizontal waveform at the junction of R42 and R43. Another check for the combined waveform can be made at the base of Q8. Its normal amplitude is approximately 1.8 V p-p, in this example.

VIDEO AMPLIFIER

Weak, distorted, or absent output from the video amplifier can be investigated initially by bringing a finger near the case of Q3 (Figure 13-11). This is a quick check; if Q3 and the following stages are workable, considerable noise interference will be observed in the reproduced raster or picture. Troubleshooting proceeds by checking for sync signal at J1; the peak voltage is normally 1 volt on open circuit, or 0.4 volt into a 75-ohm load, in this example.

Defective components or devices can usually be pinpointed by means of DC voltage measurements. Specified video-amplifier voltages for this example are listed in Table 13-1. Follow-up resistance measurements with a low-power ohmmeter are often helpful. Note that the +18 V regulated supply must be set correctly, or measured voltage values will be deceptive. If the regulated voltage is off-value, set it correctly, as explained under "Power Supply."

Open capacitors are the most difficult fault to track down, because an open capacitor does not affect the normal DC voltage and resistance values. When a capacitor is suspected of being open, the troubleshooter usually bridges it temporarily with a known good capacitor to see whether normal operation is resumed.

VHF MODULATOR

When there is normal video output, but weak or no modulated-VHF output, suspicion falls upon the VHF modulator section. (See Figure 13-12). The troubleshooter starts by checking the video-input waveform to the modulator. This waveform normally has an amplitude of approximately 1 V p-p. If it is missing, attenuated, or distorted, diodes CR1 and CR2 should be checked next. To confirm an assumption that the oscillator is not operating, the troubleshooter makes a DC voltage measurement at the junction of CR1 and CR2.

If the oscillator is operating normally, a signal-developed bias of approximately −0.38 V will be measured. Oscillator malfunction is most likely to be caused by a defective transistor. Open capacitors can present puzzling malfunctions. For example, if C2 is open, the normal oscillating frequency jumps up, and the associated TV receiver will have a blank screen. Troubleshooters usually check suspected open capacitors by temporarily bridging them with known good capacitors.

WAVEFORM DISTORTION AND NEGATIVE FEEDBACK

The two chief types of waveform distortion result from incorrect frequency response (frequency distortion), and from nonlinear circuit operation (amplitude distortion). Various forms of negative feedback are widely used to control frequency response and also to control linearity of operation. Accordingly, when a fault occurs in a negative-feedback loop, the result is characteristic

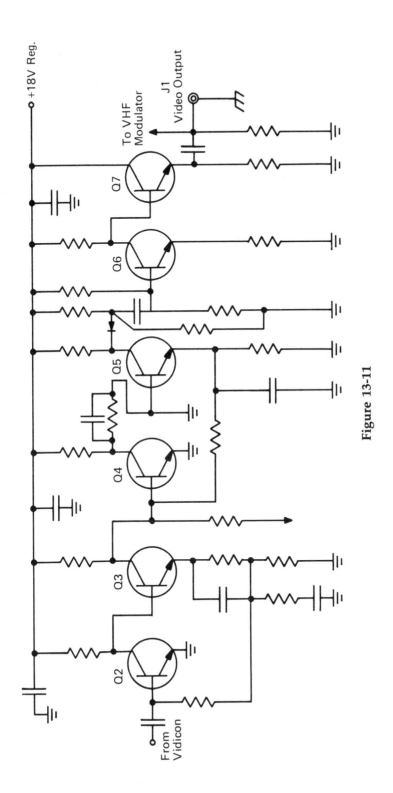

Figure 13-11

Figure 13-11 (Continued)

Note: The video input transistor is spotted easily inasmuch as it is connected to the vidicon tube output. The video output transistor is also spotted easily inasmuch as it is connected to the video-output connector.

The video amplifier has high voltage gain. An output of 1 V p-p is developed from a signal amplitude of 20 mV at the collector of Q2 (a gain of 50 times), and the signal level at the base of Q2 is not measurable with conventional service meters.

Caution: In this example, the TV camera employs a power transformer, and shock hazard is minimized. However, if you connect the video output connector to a TV receiver or a TV monitor that does not use a power transformer, a shock hazard will result accordingly.

Figure 13-11

Typical video amplifier circuitry. Rated frequency response is to 10 MHz.

Figure 13-12

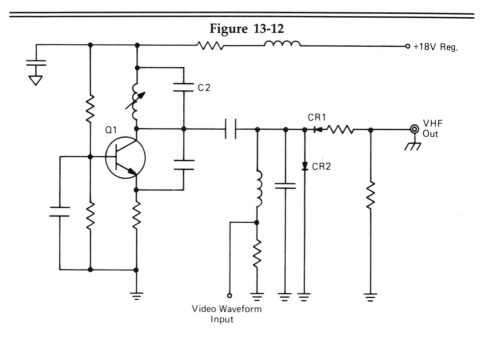

Figure 13-12 (Continued)

Note: The VHF modulator diodes and VHF oscillator transistor are spotted easily because they are connected to the VHF output connector.

Note that if a bright light is placed behind the circuit board, the printed circuit conductors are readily visible through the board.

In this example, the rated modulated-RF output is 50 millivolts into a 300-ohm load. This output level is capable of driving 300 feet of coaxial cable.

Figure 13-12

Typical VHF modulator circuitry. The RF oscillator can be adjusted for operation on a chosen channel.

waveform distortion. With reference to Figure 13-13, the four fundamental types of negative feedback are termed emitter feedback, collector feedback, frequency-selective emitter feedback, and frequency-selective collector feedback.

Some manufacturers call emitter feedback "current feedback" or "shunt feedback." A few manufacturers call emitter feedback "voltage feedback." Similarly, some manufacturers call collector feedback "voltage feedback," or "series feedback." A few manufacturers call collector feedback "current feedback." Because of this lack of consistent terminology, the troubleshooter may need to adopt his or her own preferred terminology, and to proceed on the basis of related circuit action. Sometimes, waveshaping networks include a combination of emitter feedback and collector feedback.

When "sizing up" circuit action, it is sometimes helpful to keep the following points in mind:

1. Emitter feedback reduces the stage gain in accordance with the value of the emitter resistor, and it also linearizes stage operation more or less. Observe also that emitter feedback increases the input impedance of the stage and makes it easier to drive. (Emitter feedback also increases the output impedance of the stage.)

2. Collector feedback reduces the stage gain in accordance with the value of the feedback resistor, and it also linearizes stage operation more or less. Observe also that collector feedback decreases the input impedance of the stage and makes it harder to drive. (Collector feedback also decreases the output impedance of the stage.)

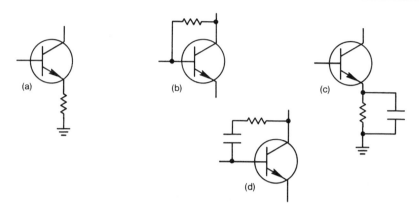

Note: Negative-feedback circuitry is widely used for waveshaping in CCTV (and other) TV-related applications. Troubleshooters know that various types of waveform distortion point to component faults in associated feedback circuitry. The feedback arrangement in (a) is called emitter feedback, or shunt feedback. The arrangement in (b) is called collector feedback, or series feedback. The arrangement in (c) is called frequency-selective emitter feedback, when the capacitor has a value that attenuates the lower frequencies in the waveform. The arrangement in (d) is called frequency-selective collector feedback, when the capacitor has a value that attenuates the lower frequencies in the waveform.

Figure 13-13

Waveform distortion and negative feedback.

3. Frequency-selective emitter feedback is also called "partially bypassed emitter" operation. It reduces stage gain with respect to the lower frequency components in the waveform, and it also linearizes stage operation more or less at lower frequencies. (The stage has lower gain at low frequencies than at high frequencies.)

4. Frequency-selective collector feedback reduces the stage gain with respect to the higher frequency components in the waveform, and it also linearizes stage operation more or less at high frequencies. (The stage has lower gain at high frequencies than at low frequencies.) Note that all forms of basic negative feedback operate to increase the total frequency range of a stage.

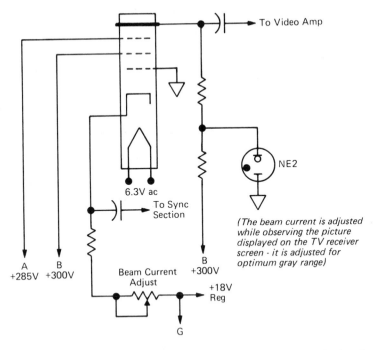

(The beam current is adjusted while observing the picture displayed on the TV receiver screen - it is adjusted for optimum gray range)

Caution: Avoid exposure of the vidicon screen to excessively bright light, such as direct sunlight—permanent screen damage can occur.

Figure 13-14

Typical vidicon circuitry. Electronic automatic beam current control is provided.

VIDICON SECTION

The 1-inch vidicon tube shown in the example of Figure 13-14 cannot reproduce a good image unless the CCTV camera lens is carefully focused and the iris adjusted for optimum contrast under the prevailing lighting conditions. If too much light is admitted into the vidicon tube, the contrast will be abnormally high and picture detail will be degraded.

On the other hand, too little light into the vidicon results in subnormal contrast and noticeable noise interference in the image. Note that a vidicon tube will be damaged immediately if it is directed toward the sun or any intense light source unless a suitable filter is used with a reduced iris opening. Note also that after a still scene has been scanned for some time, a noticeable negative after-image will persist for a while when the camera is directed at a blank wall, for example. However, this after-image will gradually decay with continued scanning action.

Troubleshooting of the vidicon circuitry starts with checking of the DC supply voltages. Observe whether the neon bulb NE2 is glowing—this gas tube regulates the target voltage in the vidicon. Follow-up resistance measurements may be made with a low-power ohmmeter. The coupling capacitors should be checked for leakage or open circuits before assuming that the vidicon tube is defective. If the vidicon is replaced, note that the focus adjustment of the camera is accomplished by repositioning the entire vidicon, yoke, and focus coil assembly.

If you want to know more about video cameras, refer to Handbook of Video Camera Servicing and Troubleshooting Techniques *by Frank Heverly, Prentice-Hall.*

Appendix A

International Frequency

Allocations

(Reproduced by permission of Prentice Hall, from *Illustrated Encyclopedic Dictionary of Electronics*, Second Edition, by John Douglas-Young)

Band No. 12:	300 gigahertz to 3 terahertz: no frequency allocated
Band No. 11:	30 to 300 gigahertz (millimetric waves), extremely high frequency (EHF)

GHz	Allocated for:
275	Amateur radio
250-275	Satellites, space research, radio astronomy
240-250	Amateur radio

Band No. 10:	3 to 30 gigahertz (centimetric waves), superhigh frequency (SHF)

GHz	Allocated For:
29.50-31.00	Fixed satellite (Earth to space)
27.50-29.50	Fixed satellite (Earth to space), fixed, mobile
25.25-27.50	Fixed, mobile
24.25-25.25	Radio navigation
24.05-24.25	Radio location, amateur radio
24.00-24.05	Amateur radio
23.60-24.00	Radio astronomy
22.00-23.60	Fixed, mobile

21.20-22.00	Earth exploration satellite (space to Earth), fixed, mobile
19.70-21.20	Fixed satellite (space to Earth)
17.70-19.70	Fixed satellite (space to Earth), fixed, mobile
15.70-17.70	Radio location
15.40-15.70	Aeronautical radio navigation
15.35-15.40	Radio astronomy
14.50-15.35	Fixed, mobile
14.40-14.50	Fixed satellite (Earth to space), fixed, mobile
14.30-14.40	Fixed satellite (Earth to space), radio navigation satellite
14.00-14.30	Fixed satellite (Earth to space), radio navigation
13.40-14.00	Radio location
13.25-13.40	Aeronautical radio navigation
12.75-13.25	Fixed, mobile
220-240	Satellites, space research, radio astronomy
200-220	Amateur radio
170-200	Satellites, space research, radio astronomy
152-170	Amateur radio
84-152	Satellites, space research, radio astronomy
71-84	Amateur radio
50-71	Satellites, space research
48-50	Amateur
40-48	Satellites
36-40	Mobile
35.2-36.0	Radio location
34.2-35.2	Radio location, space research
33.4-34.2	Radio location
32.3-33.4	Radio navigation
31.8-32.3	Radio navigation, space research
31.5-31.8	Space research
31.3-31.5	Radio astronomy
31.0-31.3	Fixed, mobile, space research
12.50-12.75	Fixed satellite (Earth to space), fixed, mobile (except aeronautical)
12.20-12.50	Broadcasting, fixed, mobile (except aeronautical)
11.70-12.20	Broadcasting, broadcasting satellite, fixed, fixed satellite (space to Earth) mobile, (except aeronautical)
11.45-11.70	Fixed satellite (space to Earth), fixed, mobile
11.20-11.45	Fixed, mobile
10.95-11.20	Fixed satellite (space to Earth), fixed, mobile
10.70-10.95	Fixed, mobile
10.68-10.70	Radio astronomy
10.60-10.68	Radio astronomy, fixed, mobile, radio location
10.55-10.60	Fixed, mobile, radio location
10.50-10.55	Radio location (CW only)

10.00-10.50	Radio location, amateur radio
9.80-10.00	Radio location, fixed
9.50-9.80	Radio location
9.30-9.50	Radio navigation, radio location
9.20-9.30	Radio location
9.00-9.20	Aeronautical radio navigation (ground-based radar), radio location
8.85-9.00	Radio location
8.75-8.85	Radio location, aeronautical radio navigation (doppler radar)
8.50-8.75	Radio location
8.40-8.50	Space research (space to Earth), fixed, mobile
5.925-8.400	Satellites, fixed, mobile
5.725-5.925	Radio location, amateur radio
5.670-5.725	Radio location, amateur radio, deep space research
5.650-5.670	Radio locatioin, amateur radio
5.470-5.650	Maritime radio navigation, radio location
5.460-5.470	Radio navigation, radio location
5.350-5.460	Aeronautical radio navigation, radio location
5.255-5.350	Radio location
5.250-5.255	Radio location, space research
5.000-5.250	Aeronautical radio navigation
4.990-5.000	Radio astronomy
4.700-4.990	Fixed, mobile
4.400-4.700	Fixed satellite (Earth to space), fixed, mobile
4.200-4.400	Aeronautical radio navigation
3.700-4.200	Fixed satellite (space to Earth), fixed, mobile
3.500-3.700	Fixed satellite (space to Earth), fixed, mobile, radio location
3.400-3.500	Fixed satellite (space to Earth), radio location, amateur
3.300-3.400	Radio location, amateur radio
3.100-3.300	Radio location

Band No. 9:	300 to 3000 megahertz (decimetric waves), ultra-high frequency (UHF)

MHz	Allocated for:
2900-3100	Radio navigation (ground-based radar), radio location
2700-2900	Aeronautical radio navigation
2690-2700	Radio astronomy
2500-2690	Satellites, fixed, mobile (except aeronautical)
2300-2500	Radio location, fixed, mobile, amateur radio (2300-2450 MHz)

2290-2300	Space research (space to Earth), fixed, mobile
1790-2290	Fixed, mobile
1770-1790	Meteorological satellite, fixed, mobile
1710-1770	Fixed, mobile
1700-1710	Space reseach (space to Earth), fixed, mobile
1690-1700	Meteorological satellite (space to Earth), meteorological aids
1670-1690	Meteorological satellite (space to Earth), meteorological aids, fixed
1660-1670	Meteorological aids, radio astronomy
1645-1660	Aeronautical mobile satellite
1644-1645	Aeronautical mobile satellite, maritime mobile satellite
1636.5-1644	Maritime mobile satellite
1558.5-1636.5	Aeronautical radio navigation
1543.5-1558.5	Aeronautical mobile satellite
1542.5-1543.5	Aeronautical mobile satellite, maritime mobile satellite
1535-1542.5	Maritime mobile satellite
1525-1535	Space operations (telemetering), Earth exploration satellite, fixed, mobile
1429-1525	Fixed, mobile
1427-1429	Space operations (telecommand), fixed, mobile (except aeronautical)
1400-1427	Radio astronomy
1350-1400	Radio location
1300-1350	Aeronautical radio navigation, radio location
1215-1300	Radio location, amateur radio
960.0-1215	Aeronautical radio navigation
942.0-960.0	Fixed
890.0-942.0	Radio location, fixed
470.0-890.0	Broadcasting (television)
460.0-470.0	Meteorological satellite, fixed, mobile (Citizens Band: 462.5375-462.7375 MHz and 467
450.0-460.0	Fixed, mobile
420.0-450.0	Radio location, amateur radio
410.0-420.0	Fixed, mobile (except aeronautical)
406.1-410.0	Radio astronomy, fixed, mobile (except aeronautical)
406.0-406.1	Mobile satellite (Earth to space)
403.0-406.0	Meteorological aids, fixed, mobile (except aeronautical)
402.0-403.0	Meteorological aids, meteorological satellite (Earth to space), fixed, mobile (except aeronautical)
401.0-402.0	Space operations (telemetering), meteorological aids, meteorological satellite (Earth to space), fixed, mobile (except aeronautical)
400.15-401.0	Space research (telemetering and tracking), meteorological satellite (maintenance telemetering), meteorological aids

400.05-400.15	Standard frequency satellite
399.9-400.05	Radio navigation satellite
335.4-399.9	Fixed, mobile
328.6-335.4	Aeronautical radio navigation (glide-path systems)

Band No. 8:	30 to 300 megahertz (metric waves), very high frequency (VHF)

MHz	Allocated for:
273.0-328.6	Fixed, mobile
267.0-273.0	Space operations (telemetering), fixed, mobile
225.0-267.0	Fixed, mobile (survival craft and equipment 243.0 MHz)
220.0-225.0	Radio location, amateur radio
216.0-220.0	Radio location, fixed, mobile
174.0-216.0	Broadcasting (television), fixed, mobile
150.05-174.0	Fixed, mobile (distress and calling-telephone-156.8 MHz)
149.9-150.05	Radio navigation satellite
148.0-149.9	Fixed, mobile
144.0-148.0	Amateur radio
138.0-144.0	Space research (space to Earth), radio location, fixed, mobile
137.0-138.0	Space research (space to Earth), space operations (telemetering and tracking), meteorological satellite
136.0-137.0	Space research (space to Earth)
117.975-136.0	Aeronautical mobile
108.0-117.975	Aeronautical radio navigation
88.0-108.0	Broadcasting (FM radio)
75.4-88.0	Broadcasting (television), fixed, mobile
74.6-75.4	Aeronautical radio navigation
73.0-74.6	Radio astronomy
54.0-73.0	Broadcasting (television), fixed, mobile
50.0-54.0	Amateur radio
38.25-50.0	Fixed, mobile
37.75-38.25	Radio astronomy, fixed, mobile
30.01-37.75	Fixed, mobile
30.005-30.01	Space operations (satellite identification), fixed, mobile

Band No. 7:	3 to 30 megahertz (decametric waves), high frequency (HF)

MHz	Allocated for:
29.70-30.005	Fixed, mobile
28.00-29.70	Amateur radio

27.50-28.00	Meteorological aids, fixed, mobile
26.10-27.50	Fixed, mobile (except aeronautical) (Citizens Band: 26.96-27.23 MHz)
25.60-26.10	Broadcasting (international AM radio)
25.11-25.60	Fixed, mobile (except aeronautical)
25.07-25.11	Maritime mobile
25.01-25.07	Fixed, mobile (except aeronautical)
24.99-25.01	Standard frequency (WWV/WWVH)
23.35-24.99	Fixed, land mobile
23.20-23.35	Aeronautical fixed and mobile
22.72-23.20	Fixed
22.00-22.72	Maritime mobile
21.87-22.00	Aeronautical fixed and mobile
21.85-21.87	Radio astronomy
21.75-21.85	Fixed
21.45-21.75	Broadcasting (international AM radio)
21.00-21.45	Amateur radio
20.01-21.00	Fixed
19.99-20.01	Standard frequency (WWV/WWVH)
18.068-19.99	Fixed
18.052-18.068	Fixed, space research
18.03-18.052	Fixed
17.90-18.03	Aeronautical mobile
17.70-17.90	Broadcasting (international AM radio)
17.36-17.70	Fixed
16.46-17.36	Maritime mobile
15.45-16.46	Fixed
15.10-15.45	Broadcasting (international AM radio)
15.01-15.10	Aeronautical mobile
14.99-15.01	Standard frequency (WWV/WWVH)
14.35-14.99	Fixed
14.00-14.35	Amateur radio
13.36-14.00	Fixed
13.20-13.36	Aeronautical mobile
12.33-13.20	Maritime mobile
11.975-12.33	Fixed
11.70-11.975	Broadcasting (international AM radio)
11.40-11.70	Fixed
11.175-11.40	Aeronautical mobile
10.10-11.175	Fixed
10.005-10.10	Aeronautical mobile
9.995-10.005	Standard frequency (WWV/WWVH)
9.775-9.995	Fixed
9.500-9.775	Broadcasting (international AM radio)

9.040-9.500	Fixed
8.815-9.040	Aeronautical mobile
8.195-8.815	Maritime mobile
7.300-8.195	Fixed
7.000-7.300	Amateur radio
6.765-7.000	Fixed
6.525-6.765	Aeronautical mobile
6.200-6.525	Maritime mobile
5.950-6.200	Broadcasting (international AM radio)
5.730-5.950	Fixed
5.450-5.730	Aeronautical mobile
5.250-5.450	Fixed, land mobile
5.060-5.250	Fixed
5.005-5.060	Fixed, broadcasting (international AM radio)
4.995-5.005	Standard frequency (WWV/WWVH)
4.850-4.995	Fixed, land mobile, broadcasting (international AM radio)
4.750-4.850	Fixed, broadcasting (international AM radio)
4.650-4.750	Aeronautical mobile
4.438-4.650	Fixed, mobile (except aeronautical)
4.063-4.438	Maritime mobile
4.000-4.063	Fixed
3.500-4.000	Fixed, mobile (except aeronautical), amateur radio
3.400-3.500	Aeronautical mobile
3.200-3.400	Fixed, mobile (except aeronautical), broadcasting (international AM radio)
3.155-3.200	Fixed, mobile (except aeronautical)
3.025-3.155	Aeronautical mobile

Band No. 6:	300 to 3000 kilohertz (hectometric waves), medium frequency (MF)

kHz	Allocated for:
2850-3025	Aeronautical mobile
2505-2850	Fixed, mobile
2495-2505	Standard frequency (WWV/WWVH)
2300-2495	Fixed, mobile, broadcasting (international AM radio)
2194-2300	Fixed, mobile
2170-2194	Mobile (distress and calling - telephone - 2182 KHz)
2107-2170	Fixed, mobile
2065-2107	Maritime mobile
2000-2065	Fixed, mobile
1800-2000	Fixed, mobile (except aeronautical), radio navigation, amateur radio

1605-1800	Fixed, mobile, aeronautical radio navigation, radio location
535-1605	Broadcasting (domestic AM radio)
525-535	Mobile, broadcasting, aeronautical radio navigation
510-525	Mobile, aeronautical radio navigation
490-510	Mobile (distress and calling - telegraph - 500 kHz)
415-490	Maritime mobile (radiotelegraphy only)
405-415	Maritime radio navigation (radio direction-finding), aeronautical radio navigation
325-405	Aeronautical radio navigation, aeronautical mobile

| Band No. 5: | 30 to 300 kilohertz (kilometric waves), low frequency (LF) |

kHz	Allocated for:
285-325	Maritime radio navigation (radio beacons), aeronautical radio navigation
200-285	Aeronautical radio navigation, aeronautical mobile
160-200	Fixed
130-160	Fixed, maritime mobile
70-130	Fixed, maritime mobile, radio navigation, radio location

| Band No. 4: | 3 to 30 kilohertz (myriametric waves), very-low frequency (VLF) |

kHz	Allocated for:
20.05-70.00	Fixed, maritime mobile (WWVB at 60 kHz)
19.95-20.05	Standard frequency (WWV/WWVH)
14.00-19.95	Fixed, maritime mobile
10.00-14.00	Radio navigation, radio location
10.00	No allocations

| Band No. 3: | 300 to 3000 hertz: no frequencies allocated |
| Band No. 2: | 30 to 300 hertz: no frequencies allocated |

Notes

1. The foregoing internationally agreed frequency allocations apply in the Western Hemisphere (Region 2). Not all allocated frequencies are actually being used.
2. The expression "fixed" means point-to-point radio communication between fixed stations.

INDEX